Biochemistry Involving Carbon-Fluorine Bonds

Biochemistry Involving Carbon–Fluorine Bonds

Robert Filler, EDITOR

Illinois Institute of Technology

A symposium sponsored by

the Divisions of Fluorine

and Biological Chemistry

at the 170th Meeting of the

American Chemical Society,

Chicago, Ill., Aug. 26, 1975.

A C S   S Y M P O S I U M   S E R I E S   **28**

AMERICAN CHEMICAL SOCIETY

WASHINGTON, D. C.   1976

SE P

CHE M

Library of Congress CIP Data

Biochemistry involving carbon–fluorine fonds.
(ACS symposium series; 28, ISSN 0097-6156)

Includes bibliographical references and index.

1. Organofluorine compounds—Physiological effect—
Congresses. I. Filler, Robert, 1923-   . II. American
Chemical Society. Division of Fluorine. III. American
Chemical Society. Division of Biological Chemistry. IV.
Series: American Chemical Society. ACS symposium series;
28.

QP981.F55B56   612'.0157   76-13037
ISBN 0-8412-0335-0   ACSMC8 28 1-215   (1976)

ACS Symposium Series

Robert F. Gould, *Series Editor*

FOREWORD

The ACS SYMPOSIUM SERIES was founded in 1974 to provide a medium for publishing symposia quickly in book form. The format of the SERIES parallels that of the continuing ADVANCES IN CHEMISTRY SERIES except that in order to save time the papers are not typeset but are reproduced as they are submitted by the authors in camera-ready form. As a further means of saving time, the papers are not edited or reviewed except by the symposium chairman, who becomes editor of the book. Papers published in the ACS SYMPOSIUM SERIES are original contributions not published elsewhere in whole or major part and include reports of research as well as reviews since symposia may embrace both types of presentation.

CONTENTS

PREFACE

In recent years there has been increasing interest and research activity in the biochemistry and medicinal applications of compounds containing the carbon–fluorine bond. The medicinal chemistry of fluorinated organic compounds was last discussed at a symposium held at the national ACS meeting in Washington, D.C. in September 1971. Simultaneously a Ciba Foundation symposium on the biochemistry and biological activities of carbon–fluorine compounds was convened in England. The lectures and incisive discussions at the Ciba symposium, in which the most prominent contributors to the field participated, were published in 1972 by Elsevier. A survey of fluorinated compounds of medicinal interest was presented in the December 1974 issue of *Chemical Technology*.

The biochemical aspects of C–F chemistry have developed rapidly since the pioneer studies in the early 1940s by Sir Rudolph Peters, who elucidated the mechanism of the toxic action of fluoroacetate by invoking the concept of "lethal synthesis." Special mention should also be made of the elegant studies since the late 1950s by Charles Heidelberger and his colleagues on the tumor-inhibitory effects of nucleotides of fluorinted pyrimidines. Important advances in the biochemistry of organofluorine compounds continue unabated and a symposium on this subject, co-sponsored by the Divisions of Fluorine and Biological Chemistry, was held at the Chicago ACS meeting in August 1975. This volume includes all ten presentations and discussions at that symposium. The subjects are of broad interest, ranging from fluorocarboxylic acids as enzymatic and metabolic probes to the use of perfluorocarbons as "artificial blood."

We hope that these reports will further stimulate those already active in the field and create a sense of excitement and enlightenment to the curious. To the newcomer we say "Welcome—don't be afraid of fluorine compounds—they're lots of fun."

Finally, I wish to express my appreciation to all of the contributors for their patience and unstinting cooperation in making this book possible.

Illinois Institute of Technology ROBERT FILLER
Chicago, Ill.
January 1976

Fluorocarboxylic Acids as Enzymatic and Metabolic Probes

ERNEST KUN

University of California, San Francisco, Department of Pharmacology, and the
Cardiovascular Research Institute, San Francisco, Calif. 94143

The development of site specific reagents necessarily
depends on the knowledge of specific biochemical reactions,
which should include the chemical structure of catalysts, substrates, modifiers, and products of enzymatic processes. Very
little is known about catalytic sites of most enzymes, therefore
comparison of substrates with analogues may be one of the experimental approaches to study this problem. It follows also that
extensive information related to analytical enzymology (i.e.,
catalytic properties of isolated enzymes, enzymatic composition
of cells) has to preceed the meaningful application of specific
probes in complex systems. This endeavor is conceptually a
synthetic one and intends to elucidate biological functions. It
is only in relatively rare cases that inhibitors of cellular
systems act in a manner predicted from in vitro enzymology. For
example, a linear multienzymatic process can be indeed regulated
by a rate limiting enzymatic component, susceptible to a specific
inhibitor. In this case the biological significance of the
linear multienzyme system can be studied with success. If acute
inhibition does not cause rapid irreversible changes in cellular
economy, sustained inhibition may trigger a variety of compen-
satory cellular processes, encompassing both intermediary and
macromolecular metabolism. Experimental pathology and toxicology
may benefit from these studies, since pathophysiology of environmental toxic effects is likely to be traced to specific initia-
ting reactions of inhibitors with specific cellular sites. This
experimental subject is at present only in its beginning stages.
In contrast to linear systems, distributive, branching, and
cyclic multienzymatic processes (1) respond in a far more complex
manner to perturbations by a specific inhibitor. Alterations of
steady state concentrations of metabolites, and subcellular changes
of topographic distribution of intermediates of metabolic pathways will occur. Most of these consequences are unpredictable
from idealized metabolic charts (2). Experimental results
obtained with relatively uncomplicated fluorocarboxylic acids
indicate that major metabolic pathways are apparently not

regulated by properties of enzymatic components alone but more
importantly by availability of substrates. The role of primary
regulatory enzymes in cellular systems, problems of molar ratios
of substrate to enzymes in biological systems, have been reviewed
(3) but the majority of biochemical texts have not incorporated
these advances at present time. The choice of F substitution of
H or OH groups in well known carboxylic acid substrates to
obtain specific inhibitors was based on straight forward chemical considerations (4). The present paper is restricted to
experimental work related to the mode of action of fluorocarboxylic acids, developed in our laboratory.

Fluoro-dicarboxylic acids

A series of chemical and enzymological experiments (5,6,7,
8,9,10,11,12,13,14,15,16,17,18,19) with mono and difluoro oxal-
acetates, fluoroglutamates, fluoroglutarate, mono and difluoro
malate, fluorolactate and malate dehydrogenases, transaminases,
glutamate and lactate dehydrogenases suggested a reasonably uniform pattern of interaction between fluorodicarboxylic acid substrates with respective enzymatic sites. It is of interest that
introduction of one or two F atoms did not change the conformation of the parent molecule to the extent that it could not be
recognized by enzymatic sites. In general, enzyme-fluoro-sub-
strate Michaelis-Menten complexes are readily formed, but in
some instances the rates of conversion to products, i.e., the
process of enzymatic catalysis itself is drastically reduced by
F-substitution. The molecular reasons for this inhibition by F-
substitution are as yet unexplored, and constitute a significant
problem of mechanistically oriented enzymology, worthy of more
detailed investigations. As would be expected F-dicarboxylic
acids behave as relatively uncomplicated linearly competitive
inhibitors with respect to the non-fluorinated substrate molecule. Kinetic analyses of the effect of fluoro-oxalacetate in
bisubstrate systems of malate dehydrogenases (19) and lactate
dehydrogenase (14) support this conclusion. A summary of experimental results is shown in Table I. Whereas monofluoro
oxalacetate is a slowly reacting substrate of MDH, difluoro substitution converts this substrate homolog to an equally good substrate of MDH to oxalacetate itself with respect to Vmax, except
difluoro-oxalacetate has a K_m of 4.0 mM (about 4.10^3 higher than
K_m of oxalacetate). Further exploration of this significant
effect of difluoro substitution may shed light on the as yet unknown molecular mechanism of catalysis of MDH. Since the primary purpose of our investigations was to obtain biological
probes, the stability of F-dicarboxylic acids were also investigated in biological systems. Despite the promising in vitro
properties of β-monofluoro oxalacetatic acid as an inhibitor of
MDH, instability prevented its application in complex systems.
As would be expected, monofluoro oxalacetate is susceptible to

Table I

SUBSTRATE AND INHIBITORY PROPERTIES OF F-CARBOXYLIC ACIDS

No.	Enzyme	F-carboxylic acid	K_m	K_i	$\frac{Ma(S)}{Ma(F-S)}$
1	Kidney MDH(mito)	β-F-oxalacetic	0.5 μM	0.5 μM	101
2	Kidney MDH(mito)	ββ'-F$_2$-oxalacetic	4.0 mM	--	1.0
3	Liver GOT(mito)	ββ'-F$_2$-oxalacetic	--	45. μM	--
4	Liver GOT(mito)	β F-oxalacetic	transaminative defluorination		
5	Liver GDH	β F-glutaric	--	330 μM	--
6	Liver GDH	α-F-glutamate(NADP$^+$)	1800 μM	710 μM	13.0
7	Liver GDH	α-F-glutamate(NAD$^+$)	640 μM	330 μM	15.0
8	Liver malic enzyme (decarboxylating)	ββ'-difluoromalate	--	300 μM	--
9	Kidney MDH	(-)-erythrofluoromalate	--	13 μM	--
10	Muscle LDH	L(+) β-fluorolactate	--	300 μM	--

LEGEND TO TABLE I: MDH = malate dehydrogenase; GOT = glutamate oxalacetate amino-transferase; GDH = glutamate dehydrogenase; LDH = lactate dehydrogenase; Ma = molecular activity; (S) = with physiological substrate; (F-S) = with fluoro-analogue

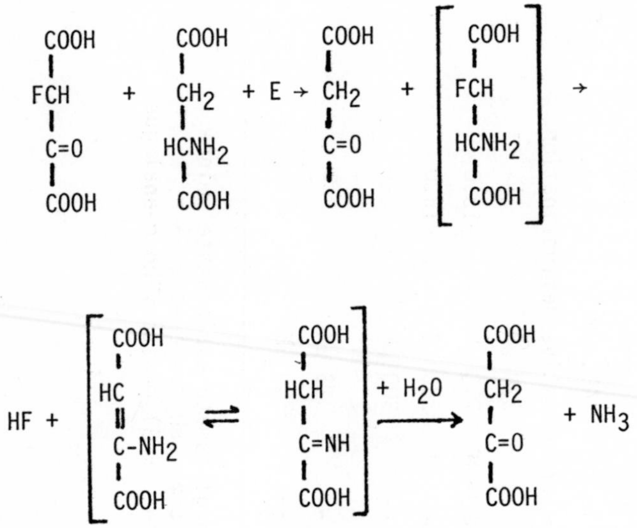

Figure 1. *Transaminative degradation of monofluoro-oxalacetic acid*

bivalent metal ion catalyzed decarboxylation to fluoro pyruvate. The kinetics of this reaction has been studied in detail (20). Interaction of monofluoro oxalacetate with glutamate-oxalacetate aminotransferase yields rapid elimination of F^- and NH_4^+ from a system containing both aspartate and the fluoro-acid (6), representing a rare case of elimination reactions characteristic of pyridoxal-phosphate catalysis, as well known from the work of Snell and his school. This is shown in Figure 1. From a combined use of fluoro-glutarate, a relatively specific inhibitor of GDH (11) and of difluoro-oxalacetate, an inhibitor of GOT (10) the regulation of both transaminative and oxidative pathways of mitochondrial glutamate metabolism was further elucidated (12). Enzymatic reduction of monofluoro-oxalacetate by MDH on a preparative scale yielded (-)erythrofluoromalic acid (21) which is an excellent inhibitor of malate dehydrogenases in vitro. In isolated hepatocytes this inhibitor acts only on cytosolic MDH isoenzyme because it does not penetrate readily through the inner mitochondrial membrane (cf. 21), therefore it is useful as a cellular probe of this MDH isoenzyme. Extensive trials with mitochondria and isolated hepatocytes - as models for the study of complex systems - indicated that only difluoro-oxalacetic and (-)erythrofluoromalic acids proved useful.   Difluoro-oxalacetate, by inhibiting GOT, proved to be stable and highly specific in its action. Mono fluoro-malate, besides inhibiting cytosolic MDH, is also an effective activator of the citrate carrier system of the inner mitochondrial membrane, as shown in Figure 2 (cf. 21). When mitochondria are incubated with citrate and the exit

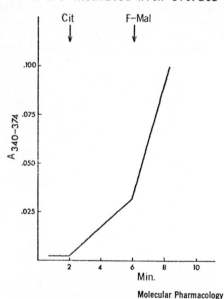

Figure 2.   *Spectrophotometric assay of citrate entry into mitochondria*

of isocitrate is monitored by an externally added $NADP^+ + Mg^{2+} +$ isocitrate dehydrogenase isocitrate detector system in the dual wave length spectrophotometer, rapid isocitrate efflux is observed under a wide variety of experimental conditions. This process is greatly stimulated by (-)erythrofluoromalate (see Figure 2) as indicated by the large increase of the rate of extramitochondrial NADPH formation. Fluoromalate appears to act similar to malate (cf. 22) except its effect is not complicated by penetration and subsequent metabolism in the matrix - as it is the case with malate. Consequently (-)erythrofluoromalate is a more specific and useful activator of the citrate-isocitrate translocating membrane system than oxidizable dicarboxylic acids (compare with 22). It is a puzzle why citrate-isocitrate exchange in energized mitochondria proceeds with the observed high efficiency (about 80% of added citrate appears as isocitrate in the extramitochondrial compartment). We presume, as discussed more extensively later, that energy dependent citrate transfer through the inner mitochondrial membrane is a more complex process than a carrier mediated carboxylic acid exchange (22) and may reflect an energy coupled membrane process of as yet unknown physiological significance.

    In vitro chemical and enzymatic studies thus far provided fluoro malate and difluoro-oxalacetate as candidates useful as probes in cellular systems. The discovery of a valuable technique, that of isolation of hepatocytes by Berry and Friend (23), gave significant impetus to further studies. Gluconeogenesis from lactate was characteristically inhibited by difluoro-oxalacetate and to a lesser extent by (-)erythrofluoromalate, whereas glucose formation from pyruvate was more sensitive to fluoromalate than to difluoro-oxalacetate (24) as shown in Table II and Figure 3. Combination of both fluoro-dicarboxylic acids

Table II

Effects of fluoro-dicarboxylic acids on rates of gluco-neogenesis in intact isolated liver cells

Cells (125—225 mg wet weight), obtained from the livers of animals fasted 18 h, were incubated for 40 min at 37 °C. The initial substrate concentration was 10 mM and inhibitor concentration 2.5 mM. The values given are means $\pm$ S.E. with the number of observations in parenthesis. Rates of glucose formation have been calculated after subtraction of the average rate observed in the absence of added substrate. This was 0.08 ± 0.003 $\mu mol \times g^{-1} \times min^{-1}$ (32 observations)

Substrate added	Inhibitor added	Glucose formed	Apparent inhibition
		$\mu mol \times g^{-1} \times min^{-1}$	%
L-Lactate	—	0.45 ± 0.02 (20)	—
L-Lactate	Fluoromalate	0.34 ± 0.06 (4)	24
L-Lactate	Difluorooxaloacetate	0.24 ± 0.04 (4)	56
L-Lactate	Fluoromalate, Difluorooxaloacetate	0.11 ± 0.05 (3)	76
Pyruvate	—	0.51 ± 0.02 (20)	—
Pyruvate	Fluoromalate	0.21 ± 0.04 (5)	59
Pyruvate	Difluorooxaloacetate	0.38 ± 0.08 (4)	25
Pyruvate	Fluoromalate, Difluorooxaloacetate	0.22 ± 0.07 (4)	57
Fructose	—	2.82 ± 0.11 (14)	—
Fructose	Fluoromalate	$2.81 \pm$ (2)	—
Fructose	Difluorooxaloacetate	$2.84 \pm$ (2)	—

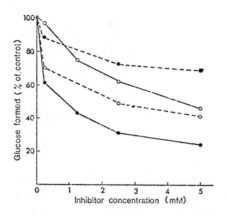

European Journal of Biochemistry

Figure 3. Effects of inhibitor concentration on rate of gluconeogenesis in isolated liver cells from pyruvate (●) or lactate (○). Each point is the average of two experiments. Conditions were the same as given in the legend of Table 1. ----, with difluorooxaloacetate; ———, with fluoromalate.

inhibit gluconeogenesis from both precursors by 76-80% without significantly altering cellular processes other than those involved in the transfer of reducing equivalents through the inner mitochondrial membrane (25, 26). In preliminary studies we found no noticable toxic effects of either fluoro carboxylic acids when injected intravenously (20 mg/100 g body weight) into mice. Marked changes in glutamate and aspartate concentrations in the liver indicated that both fluoro carboxylic acids actually penetrated liver parenchyma cells. It would appear reasonable to undertake more extensive in vivo studies with both fluoro carboxylic acids as a model for a possible experimental chemotherapeutic approach to metabolic disorders like diabetes, characterized by abnormally large gluconeogenesis.

Studies with fluorocitrate

a.)Structure of the inhibitory isomer. In contrast to fluorodicarboxylic acids of apparently low or undetectable acute toxicity,the most important fluorotricarboxylic acid: monofluorocitric acid, because of its remarkable toxicity, plays a historical significance in the field of biochemical lesions, an area established by Sir Rudolph Peters (cf. 27). In a series of classical experiments Peters and his school established that the site of action of fluorocitrate is localized at an initial step of citrate metabolism (28). The enzymatic site of action of fluorocitrate was proposed to be mitochondrial aconitase (27).

This mechanism appears to be incompatible with the results of Guarrierra-Bobyleva and Buffa (29), who found that after in vivo administration of toxic doses of fluorocitrate, only the metabolism of citrate was inhibited in mitochondria whereas cis-aconitate - which is also a substrate of aconitase - is oxidized normally. Because of the apparent uncertainties surrounding both the chemistry of the toxic isomer of fluorocitric acid and its subcellular mode of action this problem was reinvestigated.

As illustrated in Figure 4 (cf. 30) the four possible isomers of monofluorocitric acid are formed either from fluoroacetyl-CoA and oxalacetic acid or from monofluoro-oxalacetic acid and acetyl CoA. It was shown (31,32) that the toxic isomer is formed only by enzymatic condensation of fluoro acetyl CoA with oxalacetate whereas all other isomers had no significant inhibitory effect on aconitase. Because of the relatively weak inhibitory effects of fluorocitrate (K_i ≈ 50-80 μM) the distinction between "inhibitory" and "non-inhibitory" isomers was not as clear cut as one would wish for. Besides initial velocity kinetic analyses, we, as well as others, observed a time dependent, anomalous inhibition of aconitase activity, but only in the isocitrate d hydrogenase coupled test system, containing either Mg^{2+} or Mn^{2+} (see later for more details). Elucidation of the chemical structure of the toxic fluorocitrate isomer was successful despite the difficulties encountered in the enzymology of aconitase. Synthesis and resolution of isomers was accomplished in 1969 (33) and it was shown that the electrophoretically separated erythro isomers (Figure 5 and Figure 6) contained the toxic species, which was further resolved and identified as (-)erythrofluorocitric acid, correctly defined as: 1R: 2R1-fluoro-2-hydroxy-1,2,3-propane tricarboxylic acid. Crystallographic analysis of rubidium ammonium hydrogen fluorocitrate dihydrate (34) lent further support to our deduction, based originally on NMR and pk analyses of electrophoretically resolved diastereoisomers (31,32).

Since the citrate condensing enzyme plays a key role in the biosynthesis of (-)erythrofluorocitric acid from fluoroacetic acid some kinetic characteristics of this enzyme were also determined with F-acetyl CoA as substrate. Results are shown in Table III,

TABLE 3

Summary of Kinetic Properties of Citrate Synthase from Pig Heart[a]

Constant	Substrate or inhibitor	Kinetic constant
K_m	Acetyl-CoA	25 μM
K_m	Fluoroacetyl-CoA	23 μM
K_i	Fluoroacetyl-CoA	2.2 μM
V_{max}	Acetyl-CoA	2.77 (μmoles DPNH/mg/min)
V_{max}	Fluoroacetyl-CoA	0.00845 (μmoles DPNH/mg/min)

[a] DPN, malate and malate dehydrogenase served as a source of oxalacetate, and the reaction was followed by appearance of DPNH.

"The Citric Acid Cycle"

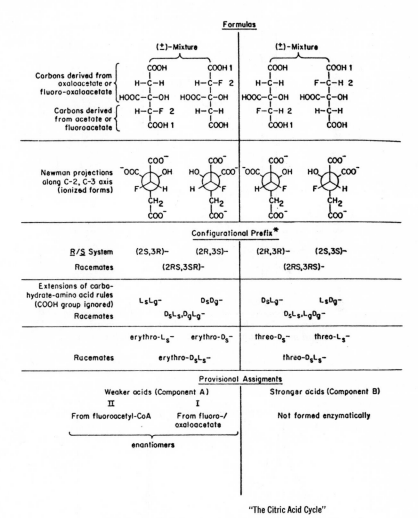

Figure 4. The isomers of monofluorocitric acid *

---

* The terminology of L, D_{c} and *erythro* D_s originates from H. B. Vickery ["A Suggested New Nomenclature for the Isomers of Isocitric Acid," *J. Biol. Chem.* 237, 1739 (1962)]. For clarification of *RS* nomenclature, the reader is referred to: Cahn, R. S., Ingold, C., and Prelog, V.: *Angew. Chem. Intern. Ed. Engl.* 385(5), 511 (1966), and Cahn, R. S., *J. Chem. Educ.* 116(41), 508 (1964).

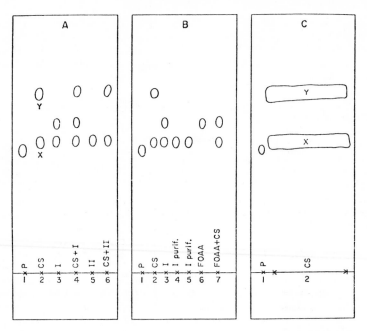

Figure 5.   Electrophoresis of fluorocitric acid from enzymatic and chemical syntheses. CS: chemically synthesized fluorocitric acid; P: picric acid marker; fluorocitric acids I and II (see text) (32).

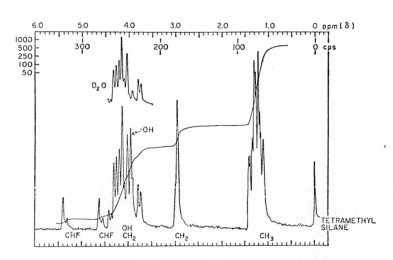

Figure 6.   Nuclear magnetic resonance spectrum of triethyl fluorocitrate (32)

(cf. 30). Whereas Km for acetyl CoA and its F-analog is the same, V_{max} in the presence of F-acetyl CoA is about 1/300th of V_{max} measured in the presence of the physiological substrate. Fluoro-acetyl CoA acts also as an inhibitor of the condensation of acetyl CoA with oxalacetate.

     b.)<u>Mode of action of (-)erythrofluorocitrate</u>. A molecular mechanism of fluorocitrate toxicity has to account for the irreversible cessation of cell function in certain specific cells of the nervous system, known to be the probable anatomical organ site of this poison (28). Whereas the question of organ specificity in the manifestation of pharmacological responses belongs to the domain of physiology, universality of metabolic enzymes in most organs still provides acceptable <u>in vitro</u> models for the study of biochemical toxicology. Consequently if inhibition of aconitase is the mode of action of fluorocitrate, irreversible inhibition of this enzyme should be demonstrated at submicromolar concentrations of the inhibitor because this range of concentration is most probably present <u>in vivo</u> (compare with slow rates of fluorocitrate biosynthesis cf. 30). Even if these requirements are met there is some uncertainty why aconitase inhibition should prove fatal, when inhibitors of other citric acid cycle enzymes causes relatively small toxicity: eg. malonate is not a powerful poison (cf. 1), or as stated earlier, inhibition of malate dehydrogenase <u>in vivo</u> elicits no detectable toxic effects. This question is of particular significance, since the existence of cytoplasmic and mitochondrial isoenzymes of aconitase (35) make it uncertain why the operation of the tricarboxylate cycle could not proceed undisturbed if only one, i.e., the mitochondrial isoenzyme, were inhibited (compare with 29). Because of the historically developed notion (36) identifying fluorocitrate toxicity with aconitase inhibition, the present evidence related to this question will be examined in detail. Notorious problems of enzyme instability at advanced stages of purification are well known to those concerned with the enzymology of aconitase. In all but one instance millimolar concentrations of ascorbate, cysteine, and Fe^{2+} are required to demonstrate full aconitase activity <u>in vitro</u>. More recent studies concerned with the mechanism of aconitase action were performed almost entirely in this highly artificial <u>in vitro</u> environment (37,38). This enzyme had a molecular weight of 68,000. Gawron et al. (39,40) while unable to reproduce the isolation procedure of Villafranca and Mildvan (37,38) obtained an enzyme (mol. weight = 66,000) which had higher specific activity than reported by earlier workers. It appears therefore doubtful that previous <u>in vitro</u> studies (37,38) are entirely valid even within the framework of cuvette enzymology. Non-competitive inhibition of heart aconitase (37) by <u>trans</u>aconitate with citrate or isocitrate as substrates (41) could not be reproduced (42), therefore the explanatory mechanism of Villafranca (41) postulating two isomeric forms of aconitase appears to be without firm experimental foundation. One of the - until recently - unrecognized complications of in vitro enzymology of aconitase lies in the frequently used NADP$^+$ dependent isocitrate dehydrogenase assay system itself. As demonstrated in 1974 (43), various preparations of aconitase are susceptible to time dependent inactivation by bivalent cations eg., Mg^{2+} or Mn^{2+}, which are necessary constituents of the coupled assay system. Unawareness of this complication can result in double reciprocal plots of reaction rates, which can simulate competitive or non-competitive or mixed types of inhibitions by fluorocitrate with highly variable apparent K_i values calculated from primary plots. We have also been misled by these types of experimental observations and postulated earlier a time dependent inactivation of aconitase by fluorocitrate, presumably by fluoro-aconitate or its defluorinated product (cf. 30). The very low K_i values for fluorocitrate reported by Brand et al. (44), assaying aconitase activities of crude mitochondrial extracts with the aid of the coupled assay, are apparently due to the inactivation of aconitase by Mg^{2+}, which is superimposed to the competitive inhibitory effect of fluorocitrate as reported more recently (43). In view of the artificiality of the enzymology of aconitase, we pursued this problem by isolating an electrophoretically homogeneous enzyme which is, for a finite period of time, fully active without any artificial activator whatsoever (43). Purification is shown in Table IV. The molecular weight of this cytoplasmic isoenzyme is

Table IV

Purification of cytoplasmic aconitase from pig liver

Purification step	Total protein	Total aconitase activity at 25°	Specific activity at 25°
	mg	μmoles/min	μmoles/min/mg protein
Step 1 (liver extract)	12,300	1,645	0.134
Step 2	560	730	1.32
Step 3	180	500	2.76
Step 4	8	85	10.6

Molecular Pharmacology

111,000, considerably higher than purified preparations requiring Fe^{2+} cysteine for activation. A much less purified mitochondrial isoenzyme has also been isolated under conditions which maintained maximal activity without added activators.

From Figure 7 it is evident that fluorocitrate acted as a linearly competitive reversible inhibitor. From suitable computer models, substrate and inhibitory constants were calculated as summarized in Table V. It is apparent that (-)erythrofluoro-

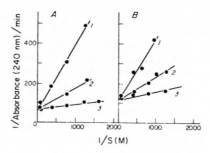

Molecular Pharmacology

Figure 7. Competitive inhibition of cytoplasmic (A) and mitochondrial (B) aconitase of pig liver by (−)-erythro-fluorocitrate. Rates of cis-aconitate formation from citrate were measured at 240 nm (10) in 0.15 M Tris-HCl, pH 7.5, at 25°. In A, 1.5 μg (protein) of cytoplasmic aconitase, and in B, 32 μg of mitochondrial aconitase, were used per test system (5-cm light path; 3-ml volume) at varied concentrations of citrate (abscissa). Curve 1, 500 μM fluorocitrate; 2, 100 μM fluorocitrate; 3, no fluorocitrate.

Table V

Substrate constants for citrate and d-isocitrate and inhibitor constants for (−)-erythro-fluorocitrate determined at pH 7.5 (10 mM Tris-HCl) and 25°

Isoenzyme	K_m			K_i		
	Citrate		d-Iso-citrate (A)	Citrate		d-Iso-citrate (A)
	A[a]	B		A	B	
	μM	μM	μM	μM	μM	μM
Cytoplasmic	220	700	17	18	27	20
Mitochon-drial	420	250	17	66	57	35

[a] A, determined by the cis-aconitate assay; B, determined by the isocitrate dehydrogenase assay (Mg²⁺ concentration varied between 0.1 and 5.3 mM).

Molecular Pharmacology

citrate behaved as an uncomplicated competitive and reversible inhibitor, therefore this kinetic property did not satisfy the mechanistic requirements set forth by the well known irreversible and highly potent toxicological effect of fluorocitrate. Although it is apparent that the enzyme requiring no arificial activator is probably closer to the enzyme functioning in the cellular environment, its properties are still different from aconitase located in isolated mitochondria. It was shown (43) that Mg^{2+} is a potent inactivator of the isolated enzyme. In sharp contrast, lysosome free mitochondria, after accumulating over 200 mM Mg^{2+} in the inner membrane and matrix (46) maintain a fully active intramitochondrial aconitase for several hours at 37°, clearly indicating that in vitro sensitivity to Mg^{2+} of isolated aconitase has no relationship to the intramitochondrial catalytic environment of this enzyme. This fact further stresses the need for extreme caution in projecting cuvette enzymology to biological systems.

Defluorination of fluorocitrate by isolated aconitase (37), has been recently reported by Villafranca and Platus (45) in the presence of an about 100 fold molar excess of fluorocitrate over aconitase. This system also contained 10 mM cysteine and 5 mM Fe^{2+}. It is noteworthy that enzyme preparations requiring no activators failed to defluorinate fluorocitrate (46), and similar results were also obtained in Cambridge (47). More recently Gawron (48) did observe in vitro defluorination of fluorocitrate, again in the presence of 5 mM Fe^{2+}, 10 mM cysteine, 30 mM ascorbate, 0.126 μ moles of aconitase and 1 mM fluorocitrate. Only 3% of 1.0 mM fluorocitrate was defluorinated and this reversible reaction stopped in 1 to 1.5 minutes. Since inhibition of the enzyme which occurs during this process is reversible (cf. 45), whereas, as shown later, mitochondrial citrate utilization is inhibited in an irreversible manner by very low concentrations of fluorocitrate, it seems unlikely that this phenomenon bears on the molecular mechanism of fluorocitrate poisoning.

In contrast to kinetic studies with isolated aconitase preparations, isolated mitochondria respond to fluorocitrate in a manner which seems to bear more directly on toxicology. When the citrate influx and isocitrate efflux (21) of isolated intact mitochondria are measured following preincubation of mitochondria with less than micromolar concentrations of (-)erythrofluorocitrate, marked and irreversible inhibition of this process is obtained. Isocitrate efflux is inhibited when either citrate or cis-aconitate are added externally (49). Results are shown in Figure 8 (a & b) and Table VI. When the mitochondrial membrane structure was disrupted by the nonionic detergent Triton X-100 full aconitase activity of mitochondria was obtained even in the presence of low concentrations of fluorocitrate, which in the same preparation completely inhibited the flux of tricarboxylic acids into intact mitochondria. It is also apparent (Figure 8b) that activation of isocitrate efflux by fluoromalate is also inhibited by preincubation with 2 μM fluorocitrate. The absence of inhibition by 10^{-8} to 10^{-6} M fluorocitrate of aconitase of lysed mitochondria agreed with in vitro enzyme kinetics (43). Fluorocitrate in intact mitochondria therefore irreversibly inhibited a membrane associated process essential for energy dependent tricarboxylic acid translocation. Similar experimental results

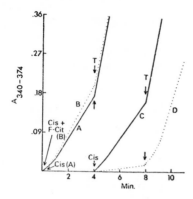

Figure 8a. Rates of isocitrate efflux from cis-aconitate
as substrate (1 mM). A = cis-aconitate alone; B =
cis-aconitate + F-citrate (5 μM) added simultaneously;
C = cis-aconitate added after 4 min preincubation
without F-citrate; D = mitochondria preincubated
with 5 μM F-citrate for 4 min, then cis-aconitate addition. At arrows, Triton X-100 is added (see Results).

Abbreviations used in Figures 8a and b: C is = cis-
aconitate; F-Cit = fluorocitrate; F-Mal = fluoromalate; Cit = citrate; T = Triton X-100.

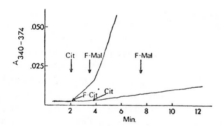

Figure 8b. Rates of isocitrate efflux from citrate (1
mM) as substrate. Upper curve = at 4 min, 1 mM
F-malate added. Lower curve = at 2 min, 2 μM F-
citrate; at 4 min, 1 mM citrate; at 7 min, 1 mM F-
malate, being added in succession.

were obtained by Brand et al. (44). Exchange diffusion of
citrate in preloaded mitochondria was not inhibited by added
fluorocitrate (44), hence it was concluded that the anion carrier
itself was not the target of inhibition by fluorocitrate. It
must be noted however, that the essential prerequisite for

Table VI

INHIBITION OF ISOCITRATE EFFLUX FROM ADDED CITRATE
AND CIS-ACONITATE BY FLUOROCITRATE

M F-citrate	% Inhibition of Isocitrate Efflux	
	citrate*	cis-aconitate*
1.0×10^{-8}	7	7
1.25×10^{-8}	11	13
2.5×10^{-8}	31	30
5.0×10^{-8}	61	51
1.0×10^{-7}	82	69
5.0×10^{-7}	100	78

*Added as substrates (1 mM) after 1 minute preincubation with F-citrate
(see Column 1).

Biochemical and Biophysical Research Communications

inhibition of mitochondrial citrate transfer is preincubation of
mitochondria for 2 to 8 minutes with very low concentrations of
fluorocitrate in the <u>absence</u> of citrate which, if added simultaneously with fluorocitrate at 5 mM concentration, prevents
inhibition. In preloaded mitochondria citrate concentration is
3-4 mM (cf. 44). It would be therefore expected that 10^{-6} to 10^{-8}
M fluorocitrate would not be effective under these conditions.
We have reinvestigated this problem in collaboration with Dr.
Eva Kirsten (visiting scientist from the University of Berlin) by
a different experimental technique. Taking advantage of the
extraordinary stability of lysosome free mitochondria (50) we
have followed the time course of ATP synthetase activity of these
organells for 40-60 minutes. Since ATP synthetase activity under
these conditions is much faster than substrate permeation into
the matrix, the time course of ^{32}P incorporation into ATP,
induced by external substrate, is a sensitive measure of the rate
of substrate translocation.

    This is illustrated in Figure 9. After preincubation of
mitochondria for 10 minutes in the presence of unlabelled P_i +
ADP to deplete endogenous substrates, ^{32}P + substrates (either
10 mM citrate - Tris, 0.5 mM malate-Tris separately or simultaneously) were added and the rate of ATP synthesis assayed (51)
by the radiochemical method. The rate of ATP synthesis reached
a plateau in 15 minutes when either substrates were added separately, but proceeded at a high rate for 60 minutes when

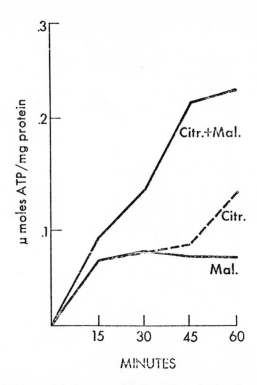

Figure 9. Time course of ATP synthesis by lysosome-free mitochondria (50). After preincubation for 10 min in the presence of 5 mM ADP + 5 mM P_i, to deplete endogenous substrates, 32 P (0.5 to 1.0 $\times$ 10⁵ dpm) and substrates (see text) were added T =30°.

citrate + malate were added simultaneously. This time course is characteristic of the kinetics of malate or fluoromalate activated citrate translocation (Figure 2). When, during preincubation of mitochondria, 5 nanomoles of (-)erythrofluorocitrate per mg mitochondrial protein and 40 mM Mg^{2+} were present and the reaction was started by either glutamate (2.5 mM) plus malate (2.5 mM) or citrate (10 mM) plus malate (0.5 mM), the rate of ATP synthesis was completely inhibited when citrate was the permeant substrate but was unaffected when glutamate + malate were the substrates (Figure 10). It is interesting that Mg^{2+} accumulation (∿230 mM into the matrix, cf. 50) caused a 3-fold augmentation of ATP synthetase activity (compare Figures 9 with 10), thus Mg^{2+} has no inhibitory effect on aconitase in situ. The critical role of the time of preincubation with 50 picomoles fluorocitrate

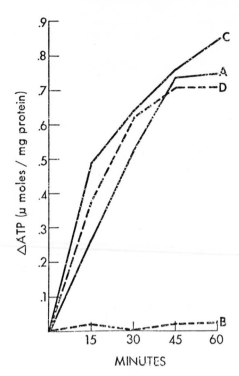

MINUTES

Figure 10. Conditions were the same as described for Figure 9, except 40 mM Mg^{2+} was also present. C = citrate + malate control; A = glutamate + malate control; D = glutamate + malate + preincubation with 50 p moles of F-citrate per mg mitochondrial protein; B = citrate + malate + preincubation with 50 p moles of F-citrate per mg mitochondrial protein; T = 30°.

per mg mitochondrial protein is illustrated by the following results. Citrate dependent ATP synthetase activity was inhibited by 52% after 3 minutes preincubation, 63% after 6 minutes preincubation, and 90-100% after 12 minutes preincubation. This time course has direct relationship to the known kinetics of fluorocitrate poisoning, which has a slow onset but is irreversible. Once citrate metabolism of mitochondria is inhibited by traces of fluorocitrate no subsequent manipulation can reactivate this specific process. Penetration of other mitochondrial substrates and their subsequent metabolism is unaffected under conditions when citrate permeation is completely inhibited. No precise molecular interpretation of these results is as yet

possible, however it seems clear that predictions based on cuvette enzymology of isolated aconitase does not fully account
for these phenomena. Technology of enzyme isolation in all but
one case modifies aconitase and makes its catalytic function dependent on artificial activators. The preparation which is fully
active without Fe^{2+} + cysteine is inactivated in vitro by Mg^{2+}
(or Mn^{2+}) yet in the intact mitochondria this inactivation by
Mg^{2+} does not take place, therefore the micro environment of the
mitochondrial enzyme has not yet been reproduced in vitro.
Inhibition of the energy dependent citrate transport by traces
of fluorocitrate in isolated mitochondria exhibits many characteristics which appear to predict the in vivo toxicity of fluorocitrate. Further analysis of this system and isolation of inner
membrane components of the transport apparatus are necessary for
real progress in this field. It is significant that Peters (cf.
27) intuitively predicted that the biochemical lesion in fluorocitrate toxicity is probably found in the realm of membrane biochemistry. Our recent results fully support this view, especially because no biochemically meaningful argument can be proposed for the critical role of citrate transfer in mitochondrial
and especially in cellular metabolism unless an as yet unknown
function of this process in the maintenance of energy transduc-
tion of the inner mitochondrial membrane is postulated.

Acknowledgement

Most of the experimental work was supported by HD-01239,
except the experiments concerned with ATP-synthetase (Figures
9 and 10) was sponsored by GM-20552. E. Kun is a research
Career Awardee of the USPH.

Literature Cited

1. Webb, J.L. "Enzyme and Metabolic Inhibitors," Acad. Press,
   New York 1963.
2. Dagley, S. and Nicholson, D.E. "An Introduction to Metabolic
   Pathways," John Wiley & Sons, Inc., New York 1970.
3. Sols, A. and Gancedo, C. "Primary Regulatory Enzymes and
   Related Proteins," in Biochemical Regulatory Mechanisms in
   Eukaryotic Cells, (Eds. Kun, E. and Grisolia, S.) Wiley
   Interscience, New York 1972.
4. Kun, E. and Dummel, R.J. "Methods in Enzymology, Vol. XIII,"
   (Eds. Colowick, Kaplan and Lowenstein) Chapt 79, p.623,
   Acad. Press, New York 1969.
5. Kun, E., Grassetti, D.R., Fanshier, D.W. and Featherstone,
   R.M. Biochem. Pharmacol. (1958) 158 (207).
6. Kun, E., Fanshier, D.W. and Grassetti, D.R. J. Biol. Chem.
   (1960) 235 (416).
7. Kun, E. and Williams-Ashman, H.G. Nature (London) (1962)

$\underline{194}$ (376).
8. Kun, E. and Williams-Ashman, H.G. Biochim. Biophys. Acta (1962) $\underline{59}$ (719).
9. Kumler, W.D., Kun, E. and Shoolery, J.N. J. Org. Chem. (1962) $\underline{27}$ (1165).
10. Kun, E., Gottwald, L.K., Fanshier, D.W. and Ayling, J.E. J. Biol. Chem. (1963) $\underline{238}$ (1456).
11. Gottwald, L.K., Ayling, J.E. and Kun, E. J. Biol. Chem.(1964) $\underline{239}$ (435).
12. Kun, E., Ayling, J.E. and Baltimore, B. J. Biol. Chem. (1964) $\underline{239}$ (2896).
13. Kun, E. and Achmatowicz, B. J. Biol. Chem. (1965) $\underline{240}$ (2619).
14. Ayling, J.E. and Kun, E. Mol. Pharmacol. (1965) $\underline{1}$ (255).
15. Gottwald, L.K. and Kun, E. J. Org. Chem. (1965) $\underline{30}$ (877).
16. Cymerman-Craig, J., Dummel, R.J., Kun, E. and Roy, S.K. Biochem. (1965) $\underline{4}$ (2547).
17. Dupourque, D. and Kun, E. Eur. J. Biochem. (1968) $\underline{6}$ (151).
18. Dupourque, D. and Kun, E. "Methods in Enzymology, Vol. XIII," (Eds. Colowick, Kaplan and Lowenstein) Chapt 20, p.116, Acad. Press, New York 1969.
19. Dupourque, D. and Kun, E. Eur. J. Biochem. (1969) $\underline{7}$ (247).
20. Dummel, R.J., Berry, M.N. and Kun, E. Mol. Pharmacol. (1971) $\underline{7}$ (367).
21. Skilleter, D.N., Dummel, R.J. and Kun, E. Mol. Pharmacol. (1972) $\underline{8}$ (139).
22. Chappell, J.B. British Med. Bull. (1968) $\underline{24}$ (150).
23. Berry, M.N. and Friend, D.S. J. Cell. Biol. (1969) $\underline{43}$ (506).
24. Berry, M.N. and Kun, E. Eur. J. Biochem. (1972) $\underline{27}$ (395).
25. Berry, M.N., Kun, E. and Werner, H.V. Eur. J. Biochem. (1973) $\underline{33}$ (407).
26. Berry, M.N., Werner, H.V. and Kun, E. Biochem. J. (1974) $\underline{140}$ (355).
27. Peters, R.A. British Med. Bull. (1969) $\underline{25}$ (223).
28. Pattison, F.L.M. and Peters, R.A. "Handbook of Experimental Pharmacology, Vol. XX," Chapt 8, p.387, Springer Press, New York 1966.
29. Guarriera-Bobyleva, V. and Buffa, P. Biochem. J. (1969) $\underline{113}$ (853).
30. Kun, E. "The Citric Acid Cycle," (Ed. Lowenstein) Chapt 6, p.297, Dekker Publications, New York 1969.
31. Fanshier, D.W., Gottwald, L.K. and Kun, E. J. Biol. Chem. (1962) $\underline{237}$ (3588).
32. Fanshier, D.W., Gottwald, L.K. and Kun, E. J. Biol. Chem. (1964) $\underline{239}$ (425).
33. Dummel, R.J. and Kun, E. J. Biol. Chem. (1969) $\underline{244}$ (2966).
34. Carrell, H.L. and Glusker, J.P. Acta Crystallogr. (1973) $\underline{29}$ (4364).
35. Eanes, R.Z. and Kun, E. Biochim. Biophys. Acta (1971) $\underline{227}$ (204).
36. Morrison, J.F. and Peters, R.A. Biochem. J. (1954) $\underline{58}$ (473).

37. Villafranca, J.J. and Mildvan, A.S. J. Biol. Chem. (1971) 246 (772).
38. Villafranca, J.J. and Mildvan, A.S. J. Biol. Chem. (1972) 247 (3454).
39. Gawron, O. Sr., Kennedy, M.C. and Bauer, R.A. Biochem. J. (1974) 143 (717).
40. Gawron, O. Sr., Waheed, A., Glaid III, A.J. and Jaklitsch, A. Biochem. J. (1974) 139 (709).
41. Villafranca, J.J. J. Biol. Chem. (1974) 249 (6149).
42. Gawron, O. Sr., manuscript submitted for publication.
43. Eanes, R.Z. and Kun, E. Mol. Pharmacol. (1974) 10 (130).
44. Brand, M.D., Evans, S.M., Mendes-Mourao, J. and Chappell, J.B. Biochem. J. (1973) 134 (217).
45. Villafranca, J.J. and Platus, E. Biochem. Biophys. Res. Comm. (1973) 55 (1197).
46. Kun, E. (unpublished experiments).
47. Peters, R.A. and Shorthouse, M. (personal communication).
48. Gawron, O. Sr. (personal communication).
49. Eanes, R.Z., Skilleter, D.N. and Kun, E. Biochem. Biophys. Res. Comm. (1972) 46 (1619).
50. Kun, E. manuscript submitted to Biochemistry.
51. Sugino, Y. and Miyoshi, Y. J. Biol. Chem. (1964) 239 (2360).

Discussion

Professor Kun

Q. What sort of methods or approaches are you using for isolating citrate carrier systems?

A. For the membrane carrier systems we are in the process of developing techniques of hydrophobic chromatography.

Q. On the affinity binding constants? You permit this substrate to remain attached to the carrier you isolated?

A. Yes.

Q. Chappell suggested that the tricarboxylate carrier is not inhibited by F-citrate. Are you suggesting that F-citrate inhibits energy coupling of the carrier?

A. The exchange diffusion carrier of Chappell was tested for inhibition by fluorocitrate (ref. 44) in citrate preloaded mitochondria under conditions (i.e., at 4 mM citrate concentrations), when also in our hands fluorocitrate inhibition is prevented by the simultaneous presence of citrate. As discussed, preincubation with very low concentrations of fluoro-citrate, in the absence of citrate are the required conditions to show inhibition of citrate translocation. Since citrate translocation in our experiments occurs in energized mitochondria (i.e., ATP dependent) it is possible that F-citrate inhibits energy coupling. No precise answer to the question is as yet available.

Q. Would you comment on the electrophoretic separation of aconitase.

A. Yes, we have done that. It is electrophoretically homogeneous.

Q. Did you use a buffer?

A. Yes, we have described this in detail in 1974 in Molecular Pharmacology and I would be glad to give you the reference (see ref. 43).

Biochemistry and Pharmacology of Ring-Fluorinated Imidazoles

KENNETH L. KIRK and LOUIS A. COHEN

Laboratory of Chemistry, National Institute of Arthritis, Metabolism, and Digestive Diseases, National Institutes of Health, Bethesda, Md. 20014

Medicinal chemists and pharmacologists have long recognized the value of fluorine in the design of analogues of metabolically significant molecules ($\underline{1}$, $\underline{2}$). The high electronegativity of fluorine can effect a marked alteration in electron-density distribution, p$\underline{K}$, conformation, etc.; simultaneously, this atom, by virtue of its small van der Waals radius (1.35 A), offers minimal steric interference to binding of the analogue at a specific macromolecular site. The effective size of fluorine attached to an sp^3 carbon may be considered to fall between those of hydrogen and the hydroxyl group, while fluorine on an sp^2 carbon is probably somewhat smaller -- the result of lone pair overlap with the adjacent pi system ($\underline{3}$, $\underline{4}$). Effective size undoubtedly varies, also, with solvation effects and specific lone pair interactions.

$$C = C - \ddot{F}: \longleftrightarrow \bar{C} - C = \overset{+}{F}:$$

Ring-fluorinated analogues of a variety of aromatic and heteroaromatic biomolecules have been synthesized and evaluated as agonists and antagonists of their natural relatives. We realized, some years ago, that the imidazole ring, one of the most ubiquitous and important of natural heteroaromatic systems, could claim no documented fluoro derivatives, and initiated an effort to fill this gap. After we had exhausted the classical synthetic routes ($\underline{5}$, $\underline{6}$), abandonment of the effort seemed inevitable. In other studies, we had found that imidazolediazonium ions, which are unusually stable to heat, are transformed photochemically to highly reactive species, possibly carbonium ions or carbenes (with expulsion of molecular nitrogen) ($\underline{7}$). It seemed reasonable that such reactive species might capture fluoride and other poorly nucleophilic anions. Indeed, the first ring-fluorinated imidazole was obtained in 1971 by photolysis of ethyl 4-diazoniumimidazole-5-carboxylate in 50% aqueous fluoroboric acid ($\underline{8}$, $\underline{9}$). Subsequently, a wide variety of both 2- and 4-fluoroimidazoles were prepared for chemical and biological studies, of which many are still in progress.

Synthetic Methods

The most general procedure for the synthesis of 4-fluoro-
imidazoles is illustrated in Figure 1 (10, 11), using aminoimidazole precursors. Unless they have the capability for resonance
overlap with an electron sink (nitro, cyano, carboxylate, etc.) at
C-5, 4-aminoimidazoles show an instability resembling that of
vinylamines, and cannot normally be isolated without partial decomposition. Catalytic hydrogenation of 4-nitroimidazoles proved
unsatisfactory as a source of nonstabilized 4-aminoimidazoles;
however, zinc dust reduction of the nitro group in 50% fluoroboric
acid, which proceeded rapidly and quantitatively at low temperature, became the method of choice. Immediately upon completion of
reduction, the aminoimidazole is diazotized with sodium nitrite in
situ, and the resulting diazonium ion is subjected to photolysis,
again in situ. After neutralization of the fluoroboric acid medium, the product is usually recovered by ethyl acetate extraction,
overall yields (based on nitroimidazole) ranging from 20-40%.
Since the other products of photolysis are generally highly polar,
hydroxylated imidazoles, they are not removed by the extraction
solvent and do not interfere with purification.

In contrast to the 4-amino series, 2-aminoimidazoles show
the stability to be expected of alkylated guanidines. These compounds are generated by catalytic hydrogenation of 2-arylazoimid-
azoles which, in turn, are obtained by coupling of the imidazole
with aryldiazonium ion (Figure 2). Although such coupling occurs
predominantly at C-2, smaller quantities of the 4- and 2,4-bis-
arylazo derivatives are formed (12). It is essential that the
desired isomer be freed of these contaminants (usually by column
chromatography) prior to hydrogenation, since purification of the
resultant 2-aminoimidazole has proved extremely laborious. The
pure 2-aminoimidazole is then diazotized and the diazonium ion is
photolyzed in situ (13).

This photochemical method has proved itself of value, with
respect to rapidity, convenience, and yield, in the synthesis of
ring-fluorinated derivatives of sensitive or complex aromatic
systems (e.g., alkaloids (14) and catecholamines (15)), as well as
other heteroaromatic systems, such as thiazole (7) and pyrrole
(16).

Special Properties

A single fluoro substituent, at C-2 or C-4, reduces the basicity of the imidazole ring by 5-6 pK units and increases its
acidity (NH $\rightarrow$ N$^-$) by 4-5 units (17). The magnitudes of these
effects are significantly greater than those of the other halogens
and indicate that, in contrast to the fluorobenzenes, the induct-
ive effect of fluorine on the imidazole ring overwhelms any
electron-releasing effect due to resonance. Inconsistencies in
the ^{19}F nmr signals for the fluoroimidazoles indicate further that

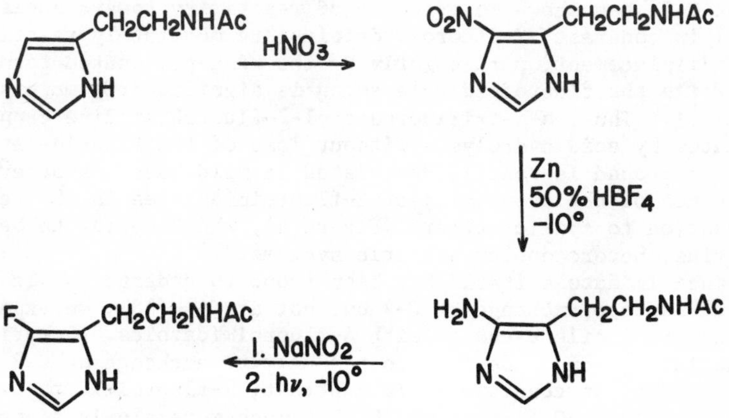

Figure 1

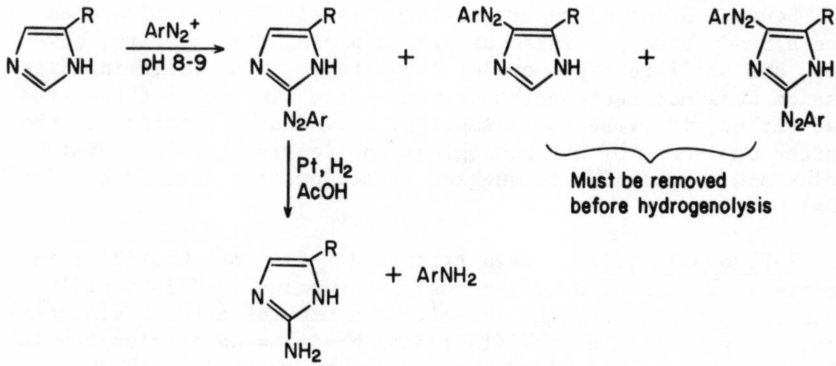

Figure 2

these compounds cannot be treated in parallel with the fluorobenzenes (17).

The fluorine atom at C-4, unless activated by an electron
sink at C-5, has shown no evidence of reactivity toward nucleo-
philes; in contrast, 2-fluoroimidazoles are moderately reactive
toward displacement, particularly in the ring-protonated form
(13), while the fluoroimidazole anion is significantly more stable
(Figure 3). Thus, α-N-trifluoroacetyl-2-fluorohistidine cannot be
deacylated by acid hydrolysis without loss of the fluorine atom,
but the compound is readily deacylated in mild base. Another
consequence of the reactivity of 2-fluoroimidazoles is the facile
condensation to cyclic trimers (Figure 4), which appear to be
large ring, heteroannular aromatic systems.

Since imidazole itself has been found to undergo facile
hydrogen-isotope exchange at C-2 but not at C-4 (18), we expected
an even more facile exchange with 4-fluoroimidazoles. Surprisingly, the latter compounds show no tendency to exchange at C-2 under
a wide variety of conditions; in contrast, 2-fluoroimidazoles exchange rapidly at C-4 above pH 10, but much more slowly at neutral
pH. Thus, a route became available for the specific tritium
labelling of 2-fluorohistidine and 2-fluorohistamine, compounds
which subsequently proved to have extensive biochemical utility.
The ability of F-2 to activate H-4 is quite unique; to date, no
other substituent at C-2, whether more or less electronegative
than fluorine, has demonstrated a capability of such magnitude.
A route to 2-[^{3}H]-4-fluorohistidine was developed recently, based
on our observation that these exchange processes are subject to
strong buffer catalysis.

Biological Properties

Several detailed reports of studies with fluoroimidazoles
have already been published or are in press; the majority, however, are still in their exploratory stages. The following discussion does not represent a comprehensive listing of these studies; rather, it presents a sampling, intended to demonstrate the
variety and scope of the biological applications of fluoroimidazoles and, hopefully, to suggest directions for further application.

2-Fluorohistidine. When tritiated 2-fluoro-L-histidine is
administered to mice subcutaneously, the amino acid is rapidly
distributed throughout the animal. Ten minutes after administration, the principal organs (including brain) show tritium levels
2-7 times that of the blood. After 72 hr, two-thirds of the
original tritium count is still within the animal, and half of
this amount is found in insoluble protein fractions (19). Since
we had shown that replacement of the 2-fluoro group by any other
substituent prevents exchange of isotope at C-4, and since all the
tritium could be back exchanged from the precipitated protein

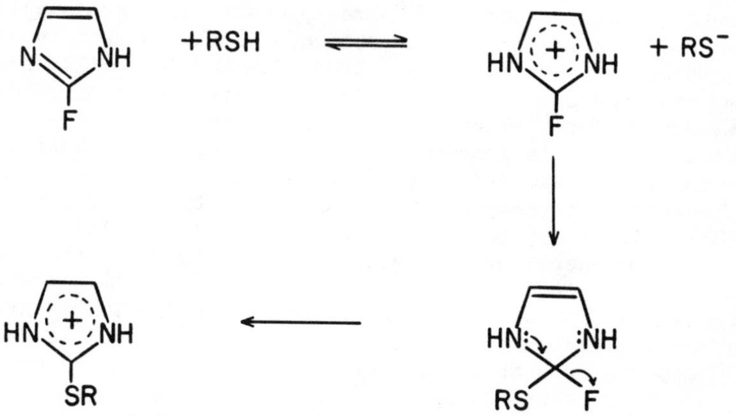

Figure 3

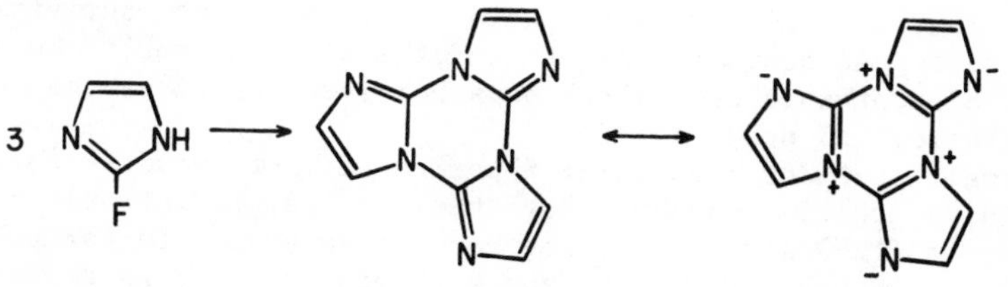

Figure 4

fractions at pH 11, the analogue must have entered into de novo
protein synthesis in place of histidine; since the 2-fluoroimid-
azole moiety is intact, covalent attachment of the fluorohistidine
must have occurred at the amino acid side chain. Such incorporation into the mouse is blocked by cycloheximide and, in isolated
liver ribosomes, by actinomycin, both drugs being inhibitors of
protein synthesis. Studies with rat pineal gland have supplied
further evidence for the incorporation of 2-fluorohistidine into
newly synthesized protein (20). Histidine and its 2-fluoro analogue are utilized competitively, since administration of an excess
of histidine reduces fluorohistidine incorporation. Presumably, the
histidine analogue is incorporated into all newly synthesized protein without discrimination; yet, it is conceivable that certain
fluorine-containing enzymes may retain activity. Single doses of
2-fluorohistidine (up to 250 mg per kilo) give no evidence of toxicity, organ degeneration, or retardation of growth of the mouse
over a 30-day period.

Concentrations of 2-fluorohistidine as low as 10^{-5} M are
bacteriostatic; inhibition of growth of E. coli (wild) is essentially complete in 3 hr at 37°, but the inhibition is reversed by
addition of histidine (20 μg/ml) (21). During the course of the inhibition, the mass of the culture increases about threefold, but
the number of viable cells increases about ninefold. These data
suggest that the bacteria, which contain two to three copies of
their chromosome during the normal growth phase, are unable to
carry out further chromosomal replication in the presence of the
drug but are able to undergo cellular division. As shown in Table
I, 2-fluorohistidine is the most effective bacteriostatic agent of
the several fluoroheterocycles tested to date.

The preceding studies suggest that 2-fluorohistidine may be
useful wherever protein synthesis in alien organisms or cells
occurs more rapidly than in the host species. Thus, the amino
acid shows antiviral behavior in various infected cell cultures
(Table II), at concentrations significantly below those needed to
produce any visible effect on the host cells (22). To our knowledge, this is the first amino acid analogue to show antiviral
properties. Although its mechanism of action has yet to be elucidated, biosynthesis of 'false' phosphorylases or virus-coat proteins are logical possibilities.

Small peptides containing fluorine are more readily obtained
by total synthesis than by biosynthesis. Thus, thyrotropin-releas-
ing factor (TRF) and luteinizing hormone-releasing factor (LHRF)
have been synthesized with 2- and 4-fluorohistidine as replace-
ments for the single histidine residues (23). While the 4-fluoro
analogues are inactive, those containing 2-fluorohistidine show
20-30% releasing activity. In view of the marked loss in basicity

Table I. Effect of Fluoro Compounds on E. coli (wild)

Compound added [a]	O.D. after 18 hr at 37° (Klett units)
None	375
2-Fluorohistidine	8
4-Fluorohistidine	375
4-Fluoroimidazole	370
2-Fluorourocanic acid	370
2-Fluoro-4-hydroxyethylthiazole	63
Same + thiamine (2 μg/ml)	270

[a] In minimal glucose medium; essentially the same results were obtained with 10^{-3}, 10^{-4}, or 10^{-5} $\underline{M}$ inhibitor.

Table II. Antiviral Activities of Fluorohistidines

Virus	Cell culture	Minimal inhibitory conc. (μg/ml) [a]		
		2-Fluoro-histidine	4-Fluoro-histidine [b]	Ribavirin [c]
VSV [d]	PRK [d]	30 [e]	100	–
HSV-1	PRK	30	>100	–
Vaccinia	PRK	30	>100	–
VSV	HeLa	30	>100	10
Coxs. B4	HeLa	30	>100	10
Polio I	HeLa	30	>100	30
Measles	VERO	10	>100	7
Coxs. B4	VERO	30	>100	30

[a] Required to reduce virus-induced cytopathogenicity by 50%.

[b] Essentially similar results were obtained for p-fluorophenylalanine. [c] An established antiviral agent, 1-β-D-ribofuranosyl-1,2,4-triazole-3-carboxamide. [d] Abbreviations are identified in ref. 32. [e] At concentrations of 100 μg/ml, morphological alteration of the host cells was apparent only after 3 days.

p-Glu-His-Pro-NH$_2$ (TRF)

p-Glu-His-Trp-Ser-Tyr-Gly-Leu-Arg-Pro-Gly-NH$_2$ (LHRF)

of the imidazole ring following introduction of fluorine, the existence of any activity is surprising, and these results suggest that recognition of the peptide by its receptor may depend more on overall conformation than on imidazole basicity.

4-Fluorohistidine. Despite the minimal steric consequences of fluorine substitution and the similarity in ring basicities of the isomers, 2-fluoro- and 4-fluorohistidine are readily differentiated by the histidine t-RNA systems. To date, the evidence indicates that 4-fluorohistidine does not substitute for histidine in protein biosynthesis, nor does this analogue show significant bacteriostatic or antiviral activity (Tables I and II). Nmr studies suggest that the isomers have different conformations (17), fluorine at C-4 being involved, perhaps, in an intramolecular hydrogen bond. Yet, the 4-fluoro isomer is a substrate for bacterial histidine decarboxylase (but not for the mammalian enzyme), while both isomers are modest substrates for histidine ammonia-lyase (Table III). The ability of 4-fluorohistidine to function both as a weak substrate and a strong competitive inhibitor for the latter enzyme prompted a reinvestigation of its mech-

Table III. Effects of Fluorohistidines on Histidine Ammonia-Lyase (pH 9, 25°)

Compound	K_m or K_1 (mM)	V_m (μM/mg/min)
L-Histidine	2.7	30
4-F-L-Histidine	1.25	0.85
2-F-L-Histidine	170	1-2

anism of action (24). Analysis of kinetic data, reversibility, isotope incorporation, and isotope effects demonstrates that the rate-limiting step for this enzyme is not the breakdown of an intermediate aminoenzyme (as previously supposed), but the almost concerted loss of a β-hydrogen atom and the covalently bound amino group. While the isomers lose ammonia at comparable rates, 4-fluorohistidine is bound much more effectively.

Urocanase, the enzyme following the ammonia-lyase in the catabolic sequence for histidine (Figure 5), promotes the hydration-rearrangement of urocanic acid (25). While neither fluorourocanic acid is significant as a substrate (Table IV), the 2-fluoro isomer

Table IV. Effects of Fluorourocanic Acids
on Urocanase (pH 7.4, 25°)

Compound	$10^7 K_m$ or K_i $(\underline{M})$	V_m (units/ml)
Urocanic acid	400	1.2
4-F-Urocanic acid [a]	-	-
2-F-Urocanic acid	1	0.008

[a]No measurable activity as inhibitor or
substrate.

is now a very potent inhibitor (26). We are not aware of any
enzyme capable of direct removal of fluorine from the imidazole
ring. The enzymes which degrade the imidazole ring of histidine,
eventually to glutamic acid, can also operate on the fluorohistidines, albeit very slowly. Thus far, we have found no evidence for
metabolites of these analogues in vivo.

As already mentioned, the fluorohistidines can be incorporated into a polypeptide sequence via total synthesis. Thus, the
peptide fragment, ribonuclease-S-(1-15), has been synthesized with
4-fluoro-L-histidine replacing histidine-12 (27). This fluorine-
containing peptide (4-F-His-RNase-(1-15)) associates with RNase-S-
(21-124) as well as, and even more strongly than, RNase-S-(1-20).
A variety of criteria suggest that the fluorohistidine-containing
aggregate has a three-dimensional structure very similar to that
of the complex, RNase-S'. The noncovalent recombination of RNase-
S-(1-15) or -(1-20) with RNase-S-(21-124) restores practically all
the enzymatic activity of the original enzyme, RNase-A; the analogous complex with 4-F-His-RNase-(1-15) is totally devoid of activity, although it is probably still capable of binding substrate.
The results provide strong support for the proposal that His-12
functions as a general acid-general base catalyst in the enzymatic
process -- a reduction of 5 units in the pK_a of the ring being
more than ample to remove this catalytic capability.

Fluorohistamines. Two types of receptor, termed H1 and H2,
have been identified as mediators of histamine's biological
actions. Characterization of these receptors, and of histamine's
biological roles, have been hampered by lack of agents which selectively bind one type. Recently, it has been found that various
substitutions at C-2 or C-4 of the histamine ring selectively decrease binding for H2 or H1 receptors, respectively. Thus, 2-
phenylhistamine retains some 20% of histamine's potency at H1
receptors, but is less than 0.1% as effective at H2 receptors.
In a preliminary study, 2-fluorohistamine was found (28) to have

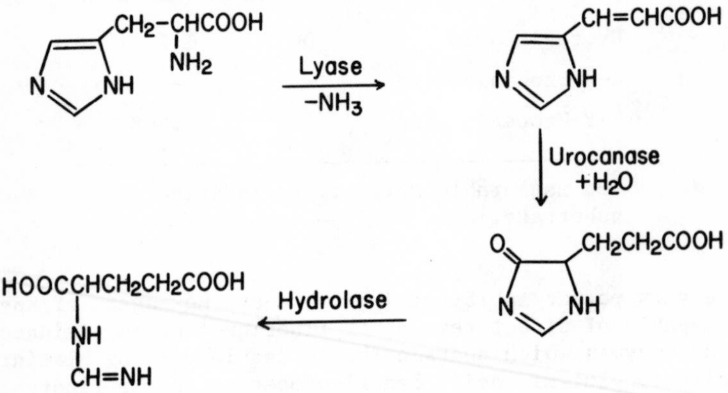

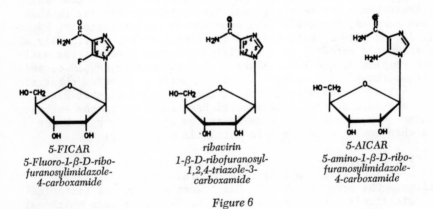

Figure 5

Figure 6

5-FICAR
5-Fluoro-1-β-D-ribo-
furanosylimidazole-
4-carboxamide

ribavirin
1-β-D-ribofuranosyl-
1,2,4-triazole-3-
carboxamide

5-AICAR
5-amino-1-β-D-ribo-
furanosylimidazole-
4-carboxamide

Figure 6

equal or greater affinity for H1 receptors (in guinea pig ileum)
than the previously most effective H1 agonist, 2-aminoethylthia-
zole. By analogy, 4-fluorohistamine might act as a selective agonist for H2 receptors, a possibility now under investigation.
2-Fluorohistamine was also found to be an effective stimulator of
cyclic AMP formation in brain slices, which contain both H1 and
H2 receptors (29).

     Fluoroimidazole Ribosides. Relatively simple imidazole
derivatives occupy key roles as intermediates in the biosynthesis
of the purine nucleosides. A major approach to the design of
antiviral agents is based on analogues of imidazole species which
might interfere with the biosynthesis of nucleosides or of nucleic
acids (Figure 6). Thus, ribavirin, a triazole analogue of 5-AICAR,
commands serious attention as a broad-spectrum antiviral agent
(30). We have synthesized 5-FICAR, a fluoro analogue of 5-AICAR
(31); this compound shows a pattern of antiviral activity quite
similar to that of ribavirin (32), and both compounds were found
to block the biosynthesis of both DNA and RNA. Ribavirin has been
shown to act by inhibiting the enzyme, IMP dehydrogenase (33), and
we assume, tentatively, that 5-FICAR functions at the same point
in the biosynthetic pathway.

Future Plans

     The results of these and other studies in fluoroimidazole
chemistry and biochemistry have raised interesting questions for
the future. Some aspects of the physical and chemical properties
of fluoroimidazoles do not conform to expectations based on data
for other imidazole systems and suggest, e.g., that some substituted imidazoles are, at best, borderline aromatic systems.
Further understanding may be provided by a study of difluoro-
imidazoles: 4,5-difluoroimidazole has been synthesized and is,
indeed, found to have anomalous properties; despite a rather extensive effort, however, the 2,4-difluoro isomer has not yet been
prepared.
     On the biological side, we may ask, e.g., why the isomeric
fluorohistidines show such marked differences in binding and response to histidine-specific enzymes; whether the replacement of
histidine by fluorohistidine in an enzyme sequence invariably
leads to loss of activity; whether the 2-fluoroimidazoles can be
made sufficiently reactive to function as covalent affinity labels
for receptor sites. Hopefully, these and other questions will
have been answered before the next Fluorine Symposium.

Literature Cited

(1) Goldman, Peter, Science (1969), 164, 1123.
(2) Ciba Foundation, "Carbon-Fluorine Compounds," Elsevier,
     Amsterdam, 1972.
(3) Reference 2, p. 211.
(4) Belsham, M. G., Muir, A. R., Kinns, Michael, Phillips,
     Lawrence, and Twanmoh, Li-Ming, J. Chem. Soc., Perkin Trans.
     2 (1974), 119.
(5) Pavlath, A. E. and Leffler, A. J., "Aromatic Fluorine Com-
     pounds," Reinhold, New York, 1962.
(6) Hudlicky, Miklos, "Organic Fluorine Chemistry," Plenum Press,
     New York, 1970.
(7) Kirk, K. L. and Cohen, L. A., unpublished observations.
(8) Kirk, K. L. and Cohen, L. A., Symposium on Fluorine in Medic-
     inal Chemistry, 162nd National Meeting of the American Chem-
     ical Society, Washington, D. C., Sept., 1971, FLUO 18.
(9) Kirk, K. L. and Cohen, L. A., J. Amer. Chem. Soc. (1971),
     93, 3060.
(10) Kirk, K. L. and Cohen, L. A., J. Amer. Chem. Soc. (1973),
     95, 4619.
(11) Kirk, K. L. and Cohen, L. A., J. Org. Chem. (1973), 38, 3647.
(12) Nagai, Wakatu, Kirk, K. L., and Cohen, L. A., J. Org. Chem.
     (1973), 38, 1971.
(13) Kirk, K. L., Nagai, Wakatu, and Cohen, L. A., J. Amer. Chem.
     Soc. (1973), 95, 8389.
(14) Lousberg, R. J. J. Ch. and Weiss, Ulrich, Experientia (1974),
     30, 1440.
(15) Kirk, K. L., manuscript in preparation.
(16) Lowenbach, W. A. and King, M. M., unpublished data.
(17) Yeh, H. J. C., Kirk, K. L., Cohen, L. A., and Cohen, J. S.,
     J. Chem. Soc., Perkin Trans. 2 (1975), 928.
(18) Wong, J. L. and Keck, J. H., Jr., J. Org. Chem. (1974), 39
     2398, and earlier references cited therein.
(19) Kirk, K. L., McNeal, Elizabeth, Cohen, L. A., and Creveling,
     C. R., manuscript in preparation.
(20) Klein, D. C., Kirk, K. L., Weller, J. L., and Parfitt, A. G.,
     Mol. Pharmacol., in press.
(21) Furano, A. V., Kirk, K. L., and Cohen, L. A., unpublished
     data.
(22) De Clercq, Erik, Kirk, K. L., and Cohen, L. A., unpublished
     data.
(23) Monahan, M. A., Vale, Wylie, Kirk, K. L., and Cohen, L. A.,
     unpublished data.
(24) Klee, C. B., Kirk, K. L., Cohen, L. A., and McPhie, Peter,
     J. Biol. Chem. (1975), 250, 5033.
(25) Kaeppeli, Franz, and Retey, Janos, Eur. J. Biochem. (1971),
     23, 198, and earlier references cited therein.
(26) Klee, C. B., Kirk, K. L., and Cohen, L. A., unpublished data.

(27) Dunn, B. M., DiBello, Carlo, Kirk, K. L., Cohen, L. A., and Chaiken, I. M., J. Biol. Chem. (1974), 249, 6295.
(28) Dismukes, R. K., unpublished data.
(29) Dismukes, R. K., Rogers, Michael, and Daly, J. W., Neurochemistry, in press.
(30) Khare, G. P., Sidwell, R. W., Witkowski, J. T., Simon, L. N., and Robins, R. K., Antimicrob. Ag. Chemother. (1973), 3, 517.
(31) Reepmeyer, J. C., Kirk, K. L., and Cohen, L. A., Tetrahedron Letters, in press.
(32) De Clercq, Erik, Luczak, Miroslav, Reepmeyer, J. C., Kirk, K. L., and Cohen, L. A., Life Sciences (1975), 17, 187.
(33) Streeter, D. G., Witkowski, J. T., Khare, G. P., Sidwell, R. W., Bauer, R. J., Robins, R. K., and Simon, L. N., Proc. Nat. Acad. Sci. USA (1973), 70, 1174.

Discussion

Q. Has there been a check of monitoring on whether C-F cleavage occurs?

A. I assume you refer to stability in biological systems. The normal pathway for histidine degradation involves conversion to urocanic acid, and, ultimately, to glutamic acid. As I had indicated, 2-fluorohistidine is a very poor substrate for the first two enzymes in this pathway; at the third step, fluoride ion would be released, together with innocuous degradation products. 4-Fluorohistidine can be transformed slowly to 4-fluorourocanic acid, which is probably a dead end. Except for this data, we have found no indications for enzymatic removal of fluorine.

Q. Have you tested the biological activity of fluoroalkylimidazoles-replacing F by CF_3?

A. We have not made any CF_3-substituted imidazoles.

Q. As far as you know, are any of them known?

A. Yes. A series of 4-trifluoromethylimidazoles have been reported by Baldwin and his colleagues at Merck Sharp & Dohme [cf. J. Med. Chem. (1975), 18, 895] and 2-trifluoromethylimidazoles by Lombardino and Wiseman [J. Med. Chem. (1974), 17, 1182.]

Q. What do you think would be the long-term stability of the C-F bond in 2-fluoroimidazoles?

A. Simple compounds, such as 2-fluoroimidazole and 2-fluoro-4-methylimidazole can be kept indefinitely in the solid state at -80°, but trimerize in a month or two at 0°. In contrast, 2-fluorohistidine appears to be indefinitely stable at room temperature. Dilute solutions of 2-fluorohistidine in phosphate buffer (pH 7) have been kept at 0° for months without evidence of deterioration.

Q.   Can you say anything about the anti-arthritic activity of the fluorohistidines?

A.   Testing has not yet been done. We are aware of the possibilities in this direction and hope to get such a study going soon.

Q.   I'd like to ask whether you ever considered the use of radioactive ^{18}F as a probe? ^{18}F has a half-life of 110 minutes.

A.   Yes, we have. Studies on the synthesis of ^{18}F imidazoles were initiated some time ago at the Atomic Energy Commission in Bucharest.

Q.   You're acquainted with the work of Al Wolf and others at Brookhaven with ^{18}F fluorophenylalanine and other biologically interesting molecules?

A.   Yes, I am. Al Wolf is aware of our activities with ^{18}F. A point of particular interest is that 2-fluorohistidine passes the blood-brain barrier very quickly, and so there is considerable interest in this compound for brain scintigraphy.

2-Fluoro-L-Histidine: A Histidine Analog Which Inhibits Enzyme Induction

DAVID C. KLEIN

Section on Physiological Controls, Laboratory of Biomedical Sciences, National
Institute of Child Health and Human Development, National
Institutes of Health, Bethesda, Md. 20014

KENNETH L. KIRK

Section on Biochemical Mechanisms, Laboratory of Chemistry, National Institute
of Arthritis, Metabolism, and Digestive Disesases, National
Institutes of Health, Bethesda, Md. 20014

Amino acid analogs have a long and interesting history (1,2) and have been extremely valuable to biochemists, pharmacologists, and clinicians as analytical tools and therapeutic agents. Some of the best established and most useful of these compounds are para-fluorophenylalanine, para-chlorophenylalanine, and alpha-methyldopa. The total number of such compounds which have been synthesized with the hopes of producing potent and specific tools probably is in the thousands.

Potential Actions of Amino Acid Analogs

The stratagem usually employed with an amino acid analog is to have it enter cells and substitute for the parent compound (1,2); the structural modification characterizing the analog precludes the normal function or metabolic fate of the parent compound. Substitution of the analog for the parent compound could involve substitution in a metabolic pathway resulting in an altered metabolite. Additionally, a metabolic pathway could be inhibited by an amino acid analog by several mechanisms, including competitive or noncompetitive inhibition of a specific enzyme or false feedback. Lastly, substitution for the parent compound in the synthesis of proteins could result in altered proteins. Such replacement of an amino acid analog for the parent compound in the primary structure of a protein could alter a protein through a number of modes of action, including alteration of the secondary, tertiary, or quarternary structure, alteration of the active site of an enzyme, and alteration of the regulatory site of an enzyme.

Role of Histidine in Enzymes

Histidine is one of the most important amino acids involved in the catalytic action of enzymes (3,4). It functions *via* the imidazole group, which can act as a general acid, general base,

or nucleophile. This catalytic role represents the generally
recognized function of histidine in enzymes (5). Some of the
enzymes in which histidine is at the active site include RNAse,
glucose-6-phosphatase, and chymotrypsin.

Recently, evidence has appeared indicating histidine may
have a further role in proteins (6). It is now generally accepted
that the activities of many enzymes are regulated via covalent
chemical modification of enzyme protein (5,7,8), including phosphorylation. Specific protein kinases catalyze phosphorylations,
and specific phosphatases catalyze dephosphorylation. As a result, the activities of enzymes can be turned on and turned off
quickly. The phosphorylation of proteins occurs via covalent
phosphorylation of specific amino acids, with the resulting
formation of both acid-stable and acid-labile phosphate bonds.
The acid-stable bonds appear to be formed with the hydroxy groups
of serine and threonine, whereas the acid-labile bonds appear to
be formed with the imidazole groups of histidine (6,8).

The importance of the imidazole group in enzyme action is
obvious; it is also possible that the imidazole group may be
important in the regulation of enzyme action. The availability
of a histidine analog identical in size and shape to histidine
but lacking the functional characteristics imparted by the
imidazole group would be of general interest.

2-Fluoro-L-Histidine: Stratagem

2-Fluoro-L-histidine was synthesized primarily to provide
an analog of histidine which would have the same size and
shape as histidine, but which would be devoid of the normal
action imparted by the imidazole group (9). (For a more detailed
discussion see L. Cohen, this Symposium Series.) Fluorine is
about the same size as the hydrogen it replaces in histidine

2-FLUORO-L-HISTIDINE

[SYNTHESIS: Kirk *ET AL.*, J. Am. Chem. Soc., *95*,8389 (1973)] and thus will not introduce major steric modifications of histidine. Fluorine-substitution, because of the strong electron withdrawing effect of fluorine, will substantially alter the pK of the imidazole group. The resulting compound will be essentially inactive, as compared to the parent compound, as a catalyst in enzymes.

This effect of fluorine substitution is demonstrated in Table I. In this experiment, the degree of transfer of a [^{14}C]-acetyl group from [^{14}C]acetyl coenzyme A to the aromatic amine tryptamine was determined by measuring the amount of N-[^{14}C]-acetyltryptamine formed. Imidazole catalyzed this transfer, whereas 2-fluoroimidazole did not. Likewise, on a theoretical basis, fluorine substitution would preclude the phosphorylation of imidazole. Thus, if 2-fluoro-L-histidine were incorporated by cells into proteins in place of histidine these proteins could not be phosphorylated at a histidine residue; nor, if the histidine were participating in the catalytic action of an enzyme, would the 2-fluoro-L-histidine-containing enzymes be active.

Pineal N-Acetyltransferase: Regulation

We have had a long standing interest in an enzyme in the pineal gland, N-acetyltransferase (10). This enzyme catalyzes the formation of N-acetylated derivatives of a number of aromatic amines (10,11,12), such as N-acetyltryptamine and N-acetylserotonin (N-acetyl-5-hydroxytryptamine). It is physiologically important because the daily 50-100 fold change in the activity of this enzyme regulates large changes in the concentration of serotonin in the pineal gland and large changes in the production of N-acetylserotonin and melatonin (5-methoxy-N-acetyltryptamine), the putative hormone of the pineal gland.

As is true of the acetyltransfer reaction presented above, it is thought that N-acetyltransferase molecules in general act through the imidazole group of histidine by attack on the thioacetyl group of acetyl CoA with the resulting formation of the unstable imidazole-acetyl bond and the subsequent transfer of the acetyl group to an acceptor amine (3). A further role of imidazole in N-acetyltransferase molecules might be to accept phosphate groups, although direct evidence for this does not exist. The increase in enzyme activity which occurs on a daily basis appears to depend not only upon protein synthesis but also, in yet an undefined way, upon the action of adenosine 3',5'-monophosphate (cyclic AMP). It is not clear, however, whether cyclic AMP activates new enzyme molecules as they are being synthesized, stimulates the formation of new enzyme molecules, or has both actions. We feel there is some evidence to suggest that enzyme activity may be regulated by an activation-inactivation mechanism. This is primarily because N-acetyltransferase activity can be rapidly destroyed, probably *via* enzyme inactivation (13,14). If an activation-inactivation mechanism plays a role in the regulation

Table I

Comparison of the catalytic activity of imidazole and 2-fluoro-
imidazole in an acetyl transfer reaction.

Reaction additions (final concentrations)	N-[^{14}C]Acetyltryptamine formed at 37°C (nanomoles/60 min incubation) *
[^{14}C]Acetyl-CoA (0.5 mM), tryptamine (10 mM)	0.14
[^{14}C]Acetyl-CoA (0.5 mM), tryptamine (10 mM), imidazole (100 mM)	13.4*
[^{14}C]Acetyl-CoA (0.5 mM), tryptamine (10 mM), 2-fluoroimidazole (100 mM)	0.20

Tubes were prepared by adding 25 µl volumes of 2 mM [^{14}C]-
acetyl-CoA (S.A. = 1 µCi/µmole), 40 mM tryptamine, 400 mM imidazole, and 400 mM 2-fluorimidazole (9), as indicated above. All
chemicals were dissolved in .1 M sodium phosphate buffer, pH 6.9.
Buffer was added to the reaction tubes to bring the final volume
to 100 µl. Data is given as the mean of 3 determinations which
were within 10% of the mean.
    *The identification of the N-[^{14}C]-acetyltryptamine extracted
into chloroform was confirmed by two dimensional TLC [chloroform,
methanol, glacial acetic acid (90:10:1) followed by ethyl ace-
tate]; 88.7% of the radioactivity applied to a pre-coated silica
gel plate, F-254 (Brinkman Instruments Co.) was isographic with
synthetic N-acetyltryptamine.
    Chloroform extraction of a tryptamine-free reaction contain-
ing 0.5 mM [^{14}C]acetyl-CoA and 100 mM imidazole yielded a small
amount of radioactivity equivalent to less than 0.02 nanomoles
of N-[^{14}C]acetyltryptamine.
    Chloroform extraction of a tryptamine-free reaction contain-
ing 100 mM 2-fluoroimidazole and 0.5 mM [^{14}C]acetyl-CoA yielded
no radioactivity.

of N-acetyltransferase, it may involve cyclic AMP. As pointed
out above, it is known that the activities of other enzymes are
regulated by cyclic AMP (7). Furthermore, the activities of some
of these enzymes are closely correlated with their phosphorylation
state. In view of our speculation that histidine phosphorylation
may be involved in the regulation of the activity of N-acetyltransferase and that histidine might be at the active site of N-
acetyltransferase, the study of the effects of analogs of histidine on the activity of N-acetyltransferase became of particular
interest.

The Effects of 2-Fluoro-L-Histidine on the Induction of Pineal
N-Acetyltransferase Activity *in vivo*.

A valuable test of the practical use of any amino acid analog is to determine if it will act in an intact normal animal
without producing severe side effects. This was one of our
initial experimental efforts with 2-fluoro-L-histidine. The
activity of N-acetyltransferase in the pineal gland can be stimulated by catecholamines acting directly on the cells containing
this enzyme (11). Predictably, the activity of this enzyme can
also be stimulated in the intact animal by the injection of these
and similar compounds, including isoproterenol (15). We found
that stimulation of N-acetyltransferase by isoproterenol was
blocked by injections of 2-fluoro-L-histidine (16).
We did not, however, detect any acute adverse side effects of
this compound, nor did we find that the injection of a similar
dose of histidine could block the induction of the enzyme.
Although this experiment showed that 2-fluoro-L-histidine could
act *in vivo*, it did not indicate whether it was acting directly
on the pineal gland, nor did it provide any other information
regarding the mode of action of this compound.

In Vitro Studies on the Induction of Pineal N-Acetyltransferase
Activity.

Early in our studies on pineal N-acetyltransferase we discovered that we could treat pineal glands in organ culture with
adrenergic agents, including adrenalin and isoproterenol, and in
this way increase the activity of N-acetyltransferase (11). This
provided an excellent model system for further studies with 2-
fluoro-L-histidine.

Effects of 2-Fluoro-L-Histidine on the Adrenergic Induction
of N-Acetyltransferase Activity. When pineal glands are treated
with 2-fluoro-L-histidine the adrenergic stimulation of N-acetyltransferase activity is substantially reduced (Table III). This
blocking effect of 2-fluoro-L-histidine was seen with glands
which were removed from animals and treated immediately with
isoproterenol and 2-fluoro-L-histidine, or with glands which had

Table II

In vivo inhibition by 2-F-HIS of the isoproterenol-induced increase in pineal N-acetyltransferase activity.

Exp	Group	Treatment	N-Acetyltransferase Activity (nmole/gland/hour)
1	A	Isoproterenol	12.6 ± 1.42
	B	Isoproterenol, 2-F-HIS	$3.0 \pm .58$
	C	Control	$0.2 \pm .011$
2	A	Isoproterenol	8.32 ± 1.24
	D	Isoproterenol, HIS	8.33 ± 1.05

Animals (100-gram male Sprague-Dawley Rats) were deprived of food overnight. Groups B and D were injected subcutaneously in the lower back region with a solution of histidine (HIS) or a suspension of 2-fluoro-L-histidine (2-F-HIS) (250 mg/kg; 25mg/0.25 ml saline) at 9:00 a.m. Groups A, B, and D were injected subcutaneously in the nape of the neck with isoproterenol (20 mg/kg, 2.0 mg/0.1 ml saline) at 10:00 a.m. Groups B and D received duplicate 2-F-HIS or HIS injections at 10:30 a.m. All animals were killed at noon and their pineal glands were removed rapidly. Values are based on 4 glands; data are presented as the mean $\pm$ S.E. (From Klein *et al.*, 16).

been chronically denervated (16). Chronic denervation removes
all neural elements from the gland; the observation that 2-fluoro-
L-histidine acted in denervated glands indicated that it was acting on pineal cells directly.

Effects of 2-Fluoro-L-Histidine on the Cyclic AMP Induction
of N-Acetyltransferase. Catecholamines stimulate the activity of
pineal N-acetyltransferase activity by first stimulating the
production of cyclic AMP (10). One possible mechanism of action
of 2-fluoro-L-histidine was inhibition of the action of cyclic
AMP. We evaluated this by determining whether 2-fluoro-L-histi-
dine could block the stimulation of N-acetyltransferase activity
as caused by a derivative of cyclic AMP, $N^6,2'$ dibutyryl adenosine 3',5'-monophosphate (dibutyryl cyclic AMP). The stimulation
of N-acetyltransferase activity caused by this compound was blocked by 2-fluoro-L-histidine (Table III). This indicated that 2-
fluoro-L-histidine was probably acting at some site in the
sequence of events which leads to an increase in enzyme activity
and follows both the interaction of catecholamines with a receptor
and the production of cyclic AMP.

Specificity of 2-Fluoro-L-Histidine Inhibition of Enzyme Induction

Our finding that the induction of pineal N-acetyltransferase
activity was blocked by 2-fluoro-L-histidine raised the question
of whether the inhibitory effects of this compound were specific
to this enzyme, or whether the induction of other enzymes could
be inhibited. To examine this important question we studied the
steroid induction of tyrosine amino transferase in fetal liver,
the benz[a]anthracene induction of aryl hydrocarbon hydroxylase
activity in liver cells, and the spontaneous increase of ornithine
decarboxylase activity in both mouse mammary tissue and the
pineal gland. In all cases the effects of 2-fluoro-L-histidine
were examined in *in vitro* systems (16).

Steroid Induction of Liver Tyrosine Amino Transferase. The
activity of tyrosine amino transferase in liver can be stimulated
by treatment with a low concentration of the steroid dexamethasone (17). We determined that the induction of tyrosine amino
transferase activity in fetal liver explants by steroid treatment
was blocked by 2-fluoro-L-histidine (Table IV, 16).

Benz[a]anthracene Induction of Liver Aryl Hydrocarbon
Hydroxylase Activity. The activity of liver aryl hydrocarbon
hydroxylase, a membrane bound enzyme, can be induced by low concentrations of a number of compounds (18), including benz[a]-
anthracene. Using a cell culture system, it was found that the
induction of the activity of this enzyme was substantially blocked
by 2-fluoro-L-histidine (Table V, 16).

Table III

Inhibition by 2-F-HIS of the drug-induced increase in pineal N-acetyltransferase activity in organ culture

Experiment	Surgical Pretreatment	Treatment in Organ Culture	N-Acetyltransferase Activity (nmole/gland/hour)
1	None	HIS (0–11 hours)	0.2 ± 0.03
	None	HIS (0–11 hours); Isoproterenol (3–11 hrs)	10.5 ± 1.5.
	None	HIS, 2-F-HIS (0–11 hours); Isoproterenol (3–11 hours)	4.9 ± 0.31 *
	None	-HIS (0–11 hours); Isoproterenol (3–11 hours)	16.4 ± 2.74
	None	-HIS, 2-F-HIS (0–11 hours); Isoproterenol (3–11 hours)	3.2 ± 0.56 *
2	None	-HIS (0–11 hours); Isoproterenol (3–11 hours)	10.4 ± 2.11
	None	-HIS, 2-F-HIS, (0–11 hours); Isoproterenol (3–11 hours)	1.6 ± 0.09 *
	SCGX	-HIS (0–11 hours); Isoproterenol (3–11 hours)	18.8 ± 3.00
	SCGX	-HIS, 2-F-HIS (0–11 hours); Isoproterenol (3–11 hours)	2.5 ± 0.52 *

Experiment	Surgical Pretreatment	Treatment in Organ Culture	N-Acetyltransferase Activity (nmole/gland/hour)
3	None	-HIS (0-11 hours); dibutyryl cyclic AMP (3-11 hrs)	22.7 ± 3.43
	None	-HIS, +2-F-HIS (0-11 hours); dibutyryl cyclic AMP (3-11 hrs)	2.6 ± 0.64*

Superior cervical ganglionectomy (SCGX) was performed 14 days prior to organ culture. Pineal glands were removed between 10:00 and 12:00 a.m. and placed into culture. The concentration of 2-fluoro-L-histidine (2-F-HIS) was 3 mM; and, when present, the concentration of histidine (HIS) was 0.1 mM. Isoproterenol was added in 5 µl of 0.01 M HCl to a final concentration of 10 µM. Dibutyryl cyclic AMP was added in 5 µl of culture medium to a final concentration of 1 mM.
Each value is based on 4 glands. Data are given as the mean ± S.E.
*Statistically less than glands treated with either isoproterenol or dibutyryl cyclic AMP
P < .01. Statistical analysis was performed using Student's "t" test. (From Klein et al., 16).

Table IV

The effect of 2-F-HIS on the dexamethasone induction of tyrosine
amino transferase activity in fetal rat liver explants in culture

Treatment in culture (hours)		Tyrosine amino transferase activity (nmole/mg protein/min)
(0-3)	(3-11)	
Control	Control	6.5
Control	Dexamethasone	54.2
2-F-HIS	2-F-HIS	6.6
2-F-HIS	2-F-HIS + dexamethasone	10.4

After a 24-hour incubation period with 1 mM histidine, the
tissue was transferred to fresh medium containing 0.1 mM histidine
for the indicated treatments. The concentration of 2-fluoro-L-
histidine (2-F-HIS) was 3 mM. Dexamethasone treatment was initiated by adding 5 µl of culture medium containing dexamethasone
resulting in a final medium concentration of 5 µM. Each value is
the average of the means of duplicate determinations performed on
duplicate cultures. The duplicate means did not differ by more
than 3 nanomoles/mg protein/minute. (From Klein et al., 16).

Spontaneous Increase of Ornithine Decarboxylase Activity.
Ornithine decarboxylase is involved in the synthesis of polyamines, which appear to play a role in the process of cell
division and gene regulation (19). The activity of this enzyme
is generally high under two circumstances: following traumatic
treatment of tissue, such as during regeneration of liver or
immediately after tissue is placed into culture, and during
normal growth of tissue.

The activity of ornithine decarboxylase increases in both
mouse mammary tissue explants (20) and in pineal glands (unpublished results, D. C. Klein and T. Oka) soon after these
tissues are placed into culture. We examined the effect of 2-
fluoro-L-histidine on the spontaneous increase of ornithine
decarboxylase activity in these systems. It was found that 2-
fluoro-L-histidine did not block the spontaneous increase of this
enzyme in the mouse mammary tissue (Table VI) but did block it in
the pineal gland (Table VII).

The lack of an effect of 2-fluoro-L-histidine on the increase
of ornithine decarboxylase in the mouse mammary tissue may be due

Table V

The effect of 2-F-HIS on the benz[a]anthracene induction of aryl
hydrocarbon hydroxylase in liver hepatoma (Hepa-1) cells

| Treatment in Culture (hours) | | Aryl hydrocarbon hydroxylase hydroxylase activity (pmoles/ mg protein/min) |
(0-3)	(3-11)	
Control	Control	19
Control	Benz[a]anthracene	159
2-F-HIS	2-F-HIS	6.2
2-F-HIS	2-F-HIS + benz[a]anthracene	21

Cells were obtained from a stock maintained in culture and
transferred to 15 X 60 mm Falcon dishes containing 3 ml of
Waymouth MAB medium with 10% fetal calf serum for 72 hours until
75-80% confluency was achieved. The medium was then replaced
with 2 ml of BGJ^b containing 0.1 mM histidine for the treatments
detailed above. The concentration of 2-fluoro-L-histidine (2-F-
HIS) was 3 mM. Benz[a]anthracene was added in a concentrated
solution resulting in a final concentration of 13 μM. Data are
presented as the means of duplicate determinations (18) performed
on duplicate cultures. The individual values did more than 5%.
(From Klein et al., 16).

Table VI

Effect of 2-F-HIS on the spontaneous increase of ornithine de-carboxylase activity in midpregnant mouse mammary explants.

Treatment	Ornithine decarboxylase activity (picomoles of $C^{14}O_2$ produced/mg tissue/hr)
Not incubated	5.0
Incubated (0-3 hours)	
Control	41.2
+ HIS (3 mM)	33.6
+ 2-F-HIS (3 mM)	33.2

Midpregnancy mouse (CH_3/Hen) mammary explants were used (20) (2-fluoro-L-histidine, 2-F-HIS; histidine, HIS). The culture medium used was M 199 ([HIS] = 0.175 mM). Data are presented as the mean enzyme activity in two cultures. Enzyme activity in each culture is based on duplicate determinations, which are within 1% of the mean. (From Klein *et al.*, 16).

Table VII

The effect of 2-F-HIS on the spontaneous increase of ornithine decarboxylase activity in pineal glands.

Treatment	N-Acetyltransferase activity (nmols/gland/hr)	Ornithine decarboxylase activity[1] (picomoles of $C^{14}O_2$ produced/gland/hr)
Not incubated	0.15 ± .02	6.38 ± 0.86
Incubated		
Control (0–11 hrs)		67.45 ± 6.47
2-F-HIS (0–11 hrs)		34.52 ± 9.5*
Control (0–3 hrs) + Isoproterenol (3–11 hrs)	13.24 ± 4.17	
2-F-HIS (0–11 hrs) + Isoproterenol (3–11 hrs)	4.37 ± 0.84	

Pineal organ cultures were prepared as described in Table II. Each datum is presented as the mean (± S.E.) of 4 to 6 determinations. [1]For the determination of ornithine decarboxylase activity, individual glands were sonicated in 100 µl of the assay buffer (20) immediately after being removed from culture. The homogenate was stored at -20C for 24 hours, thawed, and transferred to a reaction vial for enzyme assay. The final volume of the reaction was 125 µl and contained pyridoxal phosphate (40 µM), dithiothreitol (5 mM), EDTA (4 mM), Tris-HCl (50 mM), and DL-[1-C[14]]ornithine (20 µM, specific activity = 43 µCi/µmole). *Significantly lower than incubated glands not treated with 2-F-HIS, p > 0.01. (From Klein et al., 16).

to the higher concentration of histidine in the culture medium. This would block the effects of 2-fluoro-L-histidine, as discussed below. Alternatively, mammary tissue may contain more histidine or destroy 2-fluoro-L-histidine more quickly.

General Metabolic Effects of 2-Fluoro-L-Histidine

The observation that 2-fluoro-L-histidine blocked the induction of several enzymes raised the possibility that this compound, although not apparently toxic to the intact animal, was in fact highly toxic to tissue in general. To investigate this possibility we examined the effects of 2-fluoro-L-histidine on several metabolic parameters (16).
We found that after 24 hours of treatment with 2-fluoro-L-histidine, RNA synthesis was not decreased (Table VIII). This was remarkable in view of the complexity of RNA synthesis and the dependency of it upon enzyme action. In the case of protein synthesis a small inhibition due to 2-fluoro-L-histidine was apparent immediately. This inhibitory effect did not, however, increase during the next 24 hours of treatment.

Effects of 2-Fluoro-L-Histidine on Hydroxyindole-O-Methyltransferase Activity. Another topic of interest was the effect of 2-fluoro-L-histidine on the steady-state levels of a third enzyme in the pineal gland, hydroxyindole-O-methyl-transferase (10). It is generally thought that the steady state level of an enzyme depends upon both the ongoing production and destruction of enzyme molecules. Thus, even though a large increase in hydroxyindole-O-methyltransferase activity could not be studied, as in the case of the enzymes examined above, it seemed probable that the steady-state levels of this enzyme did in fact reflect enzyme production at a rate equal to that of enzyme degradation.
The effect of a 24-hour treatment with 2-fluoro-L-histidine on the activity of this enzyme, which converts N-acetylserotonin to melatonin, was examined. 2-Fluoro-L-histidine did not alter the activity of this enzyme (Table IX).

Effect of 2-Fluoro-L-Histidine on the Incorporation of Histidine into Protein. The lack of a nonspecific effect of 2-fluoro-histidine on several of the metabolic parameters examined led us to suspect that this amino acid analog may be acting on a specific mechanism, which was involved in the induction of all the enzymes we examined. Whereas it did not seem that protein synthesis was halted by 2-fluoro-L-histidine, it did seem possible that 2-fluoro-L-histidine was acting by virtue of being incorporated into protein in place of histidine.
We approached this hypothesis by first determining if the inhibition of enzyme induction by 2-fluoro-L-histidine was accompanied by an inhibition of the incorporation of histidine.

Table VIII

The effect of long term 2-F-HIS treatment in organ culture on
the incorporation of radioactivity into macromolecules in pineal
glands incubated with [^{14}C]LEU and [^{3}H]uridine for a three hour
period.

| Treatment in Organ Culture | Radioactive precursor in TCA insoluble material | |
	[^{14}C]LEU (nanomoles/gland)	[^{3}H]Uridine (picomoles/gland)
Control (0-3 hours)	0.34 ± 0.041	1.04 ± .098
2-F-HIS (0-3 hours)	0.27 ± 0.031	0.87 ± .167
Control (0-24 hours)	0.37 ± 0.018	1.14 ± .177
2-F-HIS (0-27 hours)	0.28 ± 0.018*	1.30 ± .122

Pineal glands were removed between 10:00 and 12:00 A.M. and
placed into organ culture. The culture medium contained no HIS.
The concentration of 2-F-HIS was 3 mM. [^{3}H]Uridine (3µM, S.A.=
21µCi/mole) and [^{14}C]LEU (0.38 mM, S.A.= 8.62 µCi/µmole) were
present for only the final three hours of the incubation period.
Values, which are based on 4 glands, are computed from the S.A.
of the radioactive precursors and the radioactivity per gland
precipitate. Data are presented as the means ± S.E. Statistically less than the value for control glands incubated for 27
hours (p < 0.01) but not significantly less than [^{14}C]LEU incorporation into glands treated for 3 hours with 2-F-HIS. (From
Klein *et al.*, 16).

Table IX

Lack of an effect of long-term treatment with 2-F-HIS on the
activity of hydroxyindole-0-methyltransferase.

Treatment in Organ Culture	Hydroxyindole-0-methyltransferase activity(picomoles/gland/hour)
Not incubated	69.2 $\pm$ 11.1
Control (0-24 hours)	109.2 $\pm$ 19.1
2-F-HIS (0-24 hours)	132.8 $\pm$ 8.4
Control (0-48 hours)	88.8 $\pm$ 6.0
2-F-HIS (0-48 hours)	92.8 $\pm$ 8.1

Pineal glands were removed between 10:00 and 12:00 A.M. and
placed into organ culture in medium containing 0.1 mM HIS. Each
value is based on 4 glands and is presented as the mean ($\pm$ S.E.).
The concentration of 2-F-HIS was 3 mM (from Klein *et al.*, 16).

It was found that in the presence of 7 μM [^{14}C]histidine that
[^{14}C]histidine could be incorporated into protein. More interest-
ing, however, was the finding that in the presence of 2-fluoro-
L-histidine this incorporation was blocked (Table X).

The effects of elevated extracellular concentrations of
histidine on both [^{14}C]histidine incorporation and induction
of N-acetyltransferase by isoproterenol were examined in the
presence of 2-fluoro-L-histidine (Figure 1). It was found that
the inhibitory effects of 2-fluoro-L-histidine could be overcome
by higher concentrations of histidine. This observation pointed
to the explanation that 2-fluoro-L-histidine was acting by
directly competing with histidine as a substrate in protein
synthesis.

Incorporation of [^{3}H]2-Fluoro-L-Histidine into Protein

A direct method of determining whether 2-fluoro-L-histidine
was actually being incorporated into protein was to incubate
glands with radiolabelled 2-fluoro-L-histidine and recover the
labelled amino acid from protein. We have synthesized [^{3}H]2-
fluoro-L-histidine and have found that it is incorporated into
protein and that this incorporation can be blocked by cycloheximide (an inhibitor of protein synthesis) and by high concentrations of histidine. In addition we have been able to enzymati-
cally digest labelled protein and recover a radioactive compound
which has the same chromatographic and electrophoretic character-

istics as does authentic 2-fluoro-L-histidine. This radiolabelled compound apparently is [^{3}H]2-fluoro-L-histidine (21).

This set of observations provide necessary support for the hypothesis that 2-fluoro-L-histidine is active because it is incorporated into newly synthesized protein.

Alternative Mechanisms of Action of 2-Fluoro-L-Histidine

Although the above evidence is necessary to accept the hypothesis that 2-fluoro-L-histidine is active because it is incorporated into protein, this finding might be coincidental and not causative. We have thus examined other possible modes of action of 2-fluoro-L-histidine.

Firstly, it seemed possible that 2-fluoro-L-histidine was active only after decarboxylation to 2-fluorohistamine. This, however, was proven improbable because 2-fluorohistamine itself was without effect on the adrenergic induction of N-acetyltrans-ferase activity (16). Secondly, we found that the addition of 2-fluoro-L-histidine to assays of N-acetyltransferase did not inhib-it enzyme activity, eliminating the possibility that 2-fluoro-L-histidine was a competitive or noncompetitive inhibitor of enzyme activity (16). Thirdly, we thought that 2-fluoro-L-histidine might have induced the production of an inhibitor of N-acetyl-transferase activity. We examined this possibility by adding

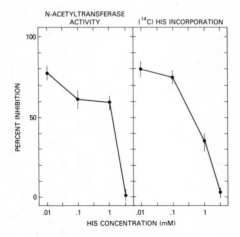

Molecular Pharmacology

Figure 1. The effect of histidine (HIS) on N-acetyltransferase activity and [^{14}C]HIS incorporation in pineal glands treated with isoproterenol and 2-fluoro-L-histidine (2-F-HIS). Data was collected from 4 experiments; a different concentration of HIS was used in each. Percent inhibition is calculated from the following: 100 [isoproterenol value—(isoproterenol + 2-F-HIS value)] ÷ isoproterenol value. In each experiment a set of 4 glands was treated with 10 μM isoproterenol and a second set of 4 glands was treated with 10 μM isoproterenol and 3 mM 2-F-HIS (16).

Table X

Effect of 2-F-HIS on radioactivity incorporated into protein of glands incubated with [³H]LEU and [¹⁴C]HIS.

Treatment in Organ Culture	Radioactivity in TCA-insoluble material (nmoles of radioactive amino acid/gland)		N-Acetyltransferase Activity (nmole/gland/hour)
	[³H]LEU	[¹⁴C]HIS	
Control (0-11 hours)	$1.25 \pm .071$	$.30 \pm .019$	$0.34 \pm .068$
2-F-HIS (0-11 hours)	$1.17 \pm .158$	$.117 \pm .018^{*}$	$0.18 \pm .034$
Control (0-3); Isoproterenol (3-11 hours)	$1.65 \pm .076$	$.41 \pm .022$	$14.3 \pm .034$
2-F-HIS (0-11 hours); Isoproterenol (3-11 hours)	$1.43 \pm .075$	$.126 \pm .005^{+}$	$3.68 \pm .548^{+}$

Pineal glands were removed between 10:00 and 12:00 and were placed into organ culture. The culture medium contained 0.38 mM [³H]LEU (S.A. = 51.2 μCi/μmole) and 1 mM [¹⁴C]HIS (S.A. = 24 μCi/μmole). The concentration of 2-F-HIS was 3 mM. Isoproterenol was added in 5 μl of 0.01 mM HCl to a final concentration of 10 μM. At the end of the experiment glands were sonicated in 100 μl of 2 mM penicillamine in 0.01 M sodium phosphate buffer, pH 6.9, at 4°C. This treatment preserves enzyme activity during handling. A 50 μl sample of each sonicate was used for enzyme assay; a 25 μl sample was used for precipitation of TCA insoluble material. Values, which are based on 4-5 glands, are computed from the S. A. of the radioactive amino acid and the radioactivity per gland precipitate.+ Data are presented as the mean ± S.E. Significantly less than control value (P < .01). [+] Significantly less than group treated with isoproterenol alone (P < .01) (From Klein et al., 16).

homogenates of glands treated with isoproterenol and 2-fluoro-L-histidine to homogenates of glands treated only with isoproterenol (16). The resulting enzyme activity was the sum of the activities of each homogenate, indicating that an inhibitor probably was not present in the isoproterenol + 2-fluoro-L-histidine homogenate. Finally, we added 2-fluoro-L-histidine to glands which had already been treated with isoproterenol for 3 hours. If 2-fluoro-L-histidine were inactivating newly formed enzyme molecules, it would have caused a rapid inactivation of enzyme activity. This, however, was not observed (16).

Conclusions

The results of our studies indicate that 2-fluoro-L-histidine certainly is not an acutely toxic compound but that it can inhibit the induction of several enzymes without blocking RNA synthesis, without inhibiting protein synthesis more than 25%, and without altering the activity of at least one pineal enzyme.

One explanation of these results is that 2-fluoro-L-histidine is incorporated into enzyme protein, and that incorporation results in the substitution of 2-fluoro-L-histidine for histidine in the primary structure of proteins. The substitution could block the induction of enzyme activity by causing changes in the structure of the enzyme, by producing non-functional active sites in the enzymes, or by blocking phosphorylation of histidine at a regulatory site on the enzyme.

We have not, however, shown whether incorporation of 2-fluoro-L-histidine into complete enzyme protein actually occurs. This is one of our future goals. We may find, alternatively, that 2-fluoro-L-histidine blocks the production of complete molecules of any protein in which it is incorporated. Net incorportation of radiolabelled leucine might appear nearly normal because an increased number of protein fragments were synthesized. It also has not been shown whether the phosphorylation of enzyme protein is blocked by the substitution of histidine with 2-fluoro-L-histidine. This and other fascinating problems regarding the action of 2-fluoro-L-histidine are presently being studied. The solutions of this problem will not only provide us with information regarding the action of 2-fluoro-L-histidine, but will also provide basic information about biological processes in general.

LITERATURE CITED

1. Shive, W. and Skinner, C. G. "Metabolic Inhibitors, A Comprehensive Treastise" pp. 1-60, Academic Press, New York (1963).

2. Fowden, L., Lewis, D., and Tristram, H., Adv. Enz. (1968) 29, 89-163.

3. Jencks, W. P. "Catalysis in Chemistry and Enzymology,"

pp 163-168, McGraw Hill, New York (1969).

4. Kirsch, J., Ann. Rev. Biochem. (1973) 42, 205-234.
5. Lehninger, "Biochemistry" pp 217-248, Worth Publishers, Inc., New York (1975).
6. Chen, C. C., Smith, D. L., Bruegger, B. B., Halpern, R. M., and Smith, R. A., Biochemistry (1974) 13, 3785-3790.
7. Segal, H. L., Science (1973) 180, 25-32.
8. Langan, T. A., Proc. Nat. Acad. Sci. USA (1968) 64, 1267-1271.
9. Kirk, K. L., Nagai, W. and Cohen, L. A., J. Am. Chem. Soc., (1973) 95, 8389-8392.
10. Klein, D. C. "The Neurosciences: Third Study Program" pp. 509-519, MIT Press, Cambridge, Mass. (1974).
11. Klein, D. C. and Weller, J. L., J. Pharmacol. Exp. Ther. 186, 516-527.
12. Deguchi, T., J. Neurochem (1975) 24(5), 1083-1085.
13. Klein, D. C. and Weller, J. L., Science (1973) 177, 532-533.
14. Binkley, S., Klein, D. C. and Weller, J. L., J. Neurochem. (1976) (in press).
15. Axelrod, J., Science (1974) 184, 1341-1348.
16. Klein, D. C., Kirk, K. L., Weller, J. L., Oka, T., Parfitt, A., and Owens, I. S., Molec. Pharmacol. (1976) (in press).
17. Kenny, F. T., "Mammalian Protein Metabolism" pp 131-145, Academic Press, New York (1970).
18. Gielen, J. E., and Nebert, D. W. J. Biol. Chem. (1972) 247, 7591-7801.
19. Russel, D. H., "Polyamines in Normal and Neoplastic Growth" Raven Press, New York (1973).
20. Oka, T. and Perry, J., J. Biol. Chem. (1976) (in press).
21. Klein, D. C., Kirk, K. L. and Weller, J. L. (manuscript in preparation).

Thymidylate Synthetase: Interaction with 5-Fluoro and 5-Trifluoromethyl-2'-Deoxyuridylic Acid

DANIEL V. SANTI, ALFONSO L. POGOLOTTI, THOMAS L. JAMES,
YUSUKE WATAYA, KATHRYN M. IVANETICH, and STELLA S. M. LAM

Department of Biochemistry and Biophysics, and Department of Pharmaceutical
Chemistry, University of California, San Francisco, Calif. 94143

Thymidylate synthetase catalyzes the reductive methylation of 2'-deoxyuridylate (dUMP) to thymidylate (TMP) with the concomitant conversion of 5, 10-methylene tetrahydrofolate ($CH_2 FAH_4$) to 7, 8-dihydrofolate (FAH_2) as depicted in Figure 1 (for a recent review, see reference 1). In this process the hydrogen at C-6 of FAH_4 is directly transferred to the methyl group of TMP (2).

One of our objectives over the past few years has been to establish the mechanism of catalysis of thymidylate synthetase. Extensive investigations of chemical counterparts (3-6) have indicated that the reaction is initiated by attack of a nucleophile at the 6-position of dUMP and that many, if not all, reactions along the pathway are facilitated by analogous nucleophilic catalysis.

The proposed mechanism of this enzyme, as derived from investigations of chemical models is illustrated in Figure 2. It is proposed that the reaction is initiated by attack of a nucleophilic group of the enzyme to the 6-position of dUMP. In this manner, the 5-position of dUMP could be made sufficiently nucleophilic (viz I, Figure 2) to react with $CH_2 FAH_4$ or an equivalent reactive species of formaldehyde. Thus, the initial condensation product between dUMP and $CH_2 FAH_4$ is now generally accepted (1, 7) to be one which is covalently bound to the enzyme and saturated across the 5, 6-double bond of dUMP (II). Proton abstraction from II would give the intermediate enolate III. As with the chemical models, III should readily undergo a β-elimination to produce the highly reactive exocyclic methylene intermediate IV and FAH_4, bound to the enzyme in close proximity. Intermolecular hydride transfer from FAH_4 to IV would yield dTMP, FAH_2, and the native enzyme. It should be emphasized that all of the aforementioned reactions and intermediates have direct chemical counterparts, and are in complete accord with all available biochemical data.

With the availability of a stable enzyme from an amethopterin resistant strain of Lactobacillus casei (8, 9) and facile methods for its purification (9-11), we undertook studies which

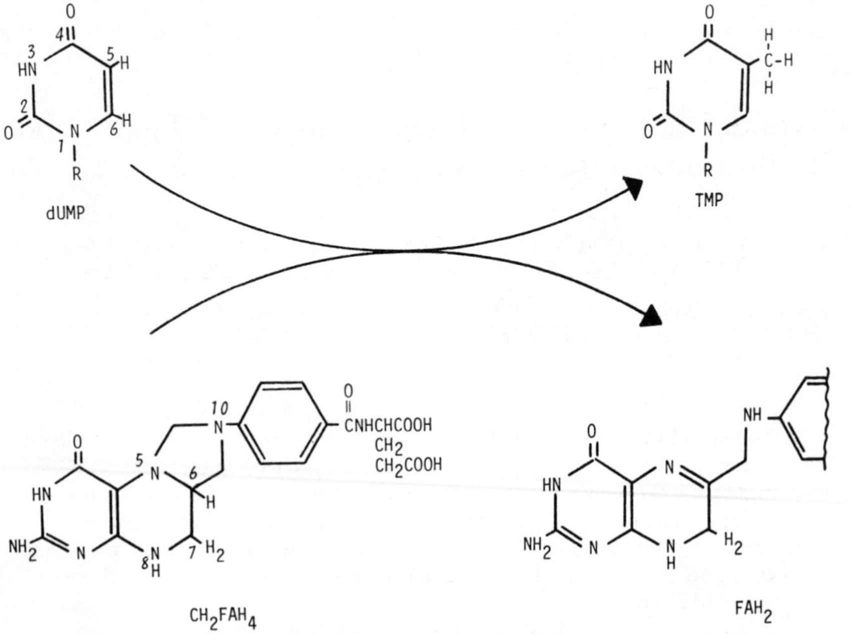

Figure 1. Reaction catalyzed by thymidylate synthetase; R = 5-phospho-2'-deoxyribosyl

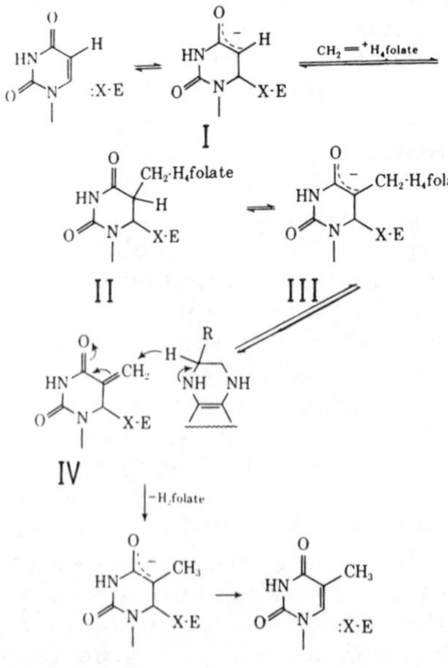

Figure 2. Suggested sequence for the thymidylate synthetase reaction. All pyrimidine structures have a 1-(5-phospho-2'-deoxyribosyl) substituent and R = CH₂NHC₆H₄COGlu.

might provide direct support for proposals which were based on the aforementioned nonenzymic models. Although a number of directions have been pursued towards this objective, the following summarizes our investigations of the interaction of thymidylate synthetase with 5-fluoro-2'-deoxyuridylate (FdUMP) and 5-trifluoromethyl-2'-deoxyuridylate (CF_3dUMP). Studies of these two fluorinated nucleotides have provided convincing evidence for the mechanism of this enzyme; in addition, they have provided insight into how these drugs act, and how the reactivity of fluorinated molecules might be utilized in the design of enzyme inhibitors. We emphasize that a number of other laboratories have been engaged in similar investigations, and regret that space does not permit complete citation of all the excellent studies performed in this area.

5-Fluoro-2'-deoxyuridylate (FdUMP). It has been known for some time (12, 13) that FdUMP is an extremely potent inhibitor of thymidylate synthetase, but the nature of inhibition has been the topic of considerable controversy (14). Since the 6-position of 1-substituted 5-fluorouracils is quite susceptible toward nucleophilic attack (15-17), we suspected that FdUMP might exert its inhibitory effect by reaction with the proposed nucleophilic catalyst of thymidylate synthetase. Studies from this (18, 19) and other (20, 21) laboratories have since demonstrated this to be the case.

A simplified depiction of the interactions of FdUMP and CH_2FAH_4 is given below in Figure 3.

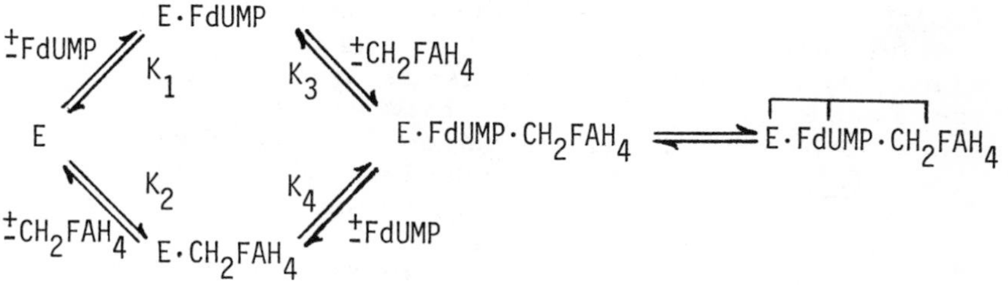

Figure 3

Using the isotope trapping method ($\underline{22}$), we have found that the two ligands, FdUMP and $CH_2 FAH_4$, interact with the enzyme in a random fashion, and that formation of the initial ternary complex ($E \cdot FdUMP \cdot CH_2 FAH_4$) is at least partially rate determining. From equilibrium binding techniques we have ascertained that the dissociation constants of both binary complexes (K_1 and K_2) are approximately $10^{-5} \underline{M}$. The first ternary complex which is formed with FdUMP, $\overline{C}H_2 FAH_4$ and enzyme is depicted as $E \cdot FdUMP \cdot CH_2 FAH_4$ and does $\underline{not}$ involve covalent bonds. Interestingly, analogous reversible ternary complexes may be formed using analogs of the cofactor. Through studies of the interaction of FdUMP and analogs of $CH_2 FAH_4$ we have ascertained that there is a striking synergism in binding of ligands to this protein. That is, the affinity of either ligand for the cognate binary complex is $\underline{ca}$. two orders of magnitude greater than the affinity for the free enzyme. With many analogs of $CH_2 FAH_4$, this ternary complex is sufficiently stable that it may be physically separated from other species ($\underline{19}$). These ternary complexes show interesting ultraviolet difference spectra which may be used for their quantitation and characterization. Shown in Figure 4 are difference spectra obtained with $CH_2 FAH_4$ and 5, 8-deazafolic acid; similar difference spectra have also been obtained with a number of other analogs of folic acid. Characteristic of these difference spectra is a peak at $\underline{ca}$. 330 nm and, usually, a trough at $\underline{ca}$. 290 nm. Although the exact reason for the spectral changes which occur upon formation of the reversible ternary complexes is yet unknown, we suggest that they result from perturbations of the p-aminobenzoylglutamate moiety of the cofactor analogs which result from their environment within the ternary complex. This perturbation is believed to be a manifestation of a conformational change which is related to the aforementioned synergism in binding of the two ligands. The difference spectrum of the $E \cdot FdUMP \cdot CH_2 FAH_4$ complex (Figure 4) is similar to those observed for the cofactor analogs, suggesting that similar changes in environment occur with the natural cofactor, $CH_2 FAH_4$. There is one striking difference in that there is a loss of differential absorbance at 269 nm in the complex formed with $CH_2 FAH_4$ ($\underline{18}$, $\underline{19}$) which we have not observed in complexes formed with cofactor analogs; the reason for this will become apparent in the ensuing discussion.

The complex formed with thymidylate synthetase, FdUMP, and the natural cofactor $CH_2 FAH_4$, has been extensively investigated. This complex is extremely tight and may readily be isolated by a variety of techniques ($\underline{18}$, $\underline{19}$, $\underline{21}$, $\underline{23}$, $\underline{24}$). Using radioactive FdUMP and $CH_2 FAH_4$, the enzyme has been titrated and shown to possess two FdUMP and two cofactor binding sites per mole ($\underline{19}$). The stoichiometry of binding has been verified in a number of laboratories by a variety of methods ($\underline{11}$, $\underline{25}$, $\underline{26}$). This is in accord with the earlier finding that thymidylate synthetase from $\underline{L}$. casei has two apparently identical subunits of MW 35,000 each (9, 27).

Studies of the rate of association of FdUMP with the $E \cdot CH_2 FAH_4$ complex (k_{on}) and its dissociation (k_{off}) have

$$E \cdot CH_2 FAH_4 + [\,^3H]FdUMP \;\; \underset{k_{off}}{\overset{k_{on}}{\rightleftharpoons}} \;\; E \cdot CH_2 FAH_4 \cdot [\,^3H]FdUMP$$

allowed us to calculate the dissociation constant of the complex to be ca. 10^{-13} $\underline{M}$. This provides an explanation for the discrepancies in K_d values reported for FdUMP in the literature; namely, previous experiments were using concentrations of enzyme higher than the K_d, and were in concentration ranges where FdUMP was behaving as a stoichiometric inhibitor. The kinetically determined K_d is approximately 10^8-fold lower than that for the binary complex; in effect, the presence of the cofactor increases the affinity of the enzyme for FdUMP by over 10 kcal/mol in binding energy. Clearly, a most pertinent feature of the interaction of FdUMP and thymidylate synthetase involves changes which occur within the bound ternary complex.

Several lines of evidence demonstrate rather conclusively that a covalent bond is formed between FdUMP and thymidylate synthetase within the complex. (a) The $E \cdot [\,^3H]FdUMP \cdot CH_2 FAH_4$ complex may be treated with a number of protein denaturants (urea, guanidine hydrochloride, etc.) without apparent loss of protein-bound radioactivity. With few exceptions, such treatment is sufficient to disrupt noncovalent interactions between low molecular weight ligands and their protein receptors. (b) Upon formation of the complex, there is a decrease of absorbance at 269 nm which corresponds to stoichiometric loss of the pyrimidine chromophore of FdUMP. This result strongly suggests that the 5,6-double bond of the pyrimidine is saturated in the bound complex. (c) The rate of dissociation of $[6-^3H]FdUMP$ from the complex shows a secondary tritium isotope effect (k_H/k_T) of 1.23. This would correspond to $k_H/k_D = 1.15$ and clearly demonstrates that the 6-carbon of the heterocycle undergoes sp^3 to sp^2 rehybridization during the process as required if the 5,6-double bond of FdUMP is saturated in the complex. (d) Proteolytic digestion of the complex yields a peptide which is covalently bound to <u>both</u> FdUMP and $CH_2 FAH_4$. The ultraviolet and fluorescence spectra of this peptide are characteristic of 5-alkyltetrahydrofolates and, as with the native complex, there is no evidence of ultraviolet absorption of the FdUMP chromophore.

From these lines of evidence, together with information gathered from model chemical counterparts, the structure of the enzyme·FdUMP·CH_2FAH_4 complex is currently believed to be as depicted in Figure 5. Here, a nucleophile of the enzyme has added to the 6-position of FdUMP, and the 5-position of the pyrimidine is coupled to the 5-position of FAH_4 <u>via</u> the methylene

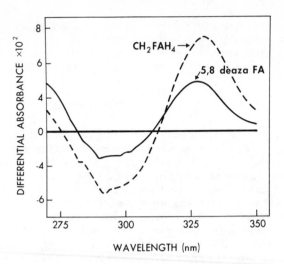

Figure 4. *Dashed line: Ultraviolet difference spectra of FdUMP, CH_2FAH_4 and thymidylate synthetase vs. CH_2FAH_4 and thymidylate synthetase. Solid line: FdUMP, 5,8-deazofolate and thymidylate synthetase vs. enzyme and 5,8-deazafolate.*

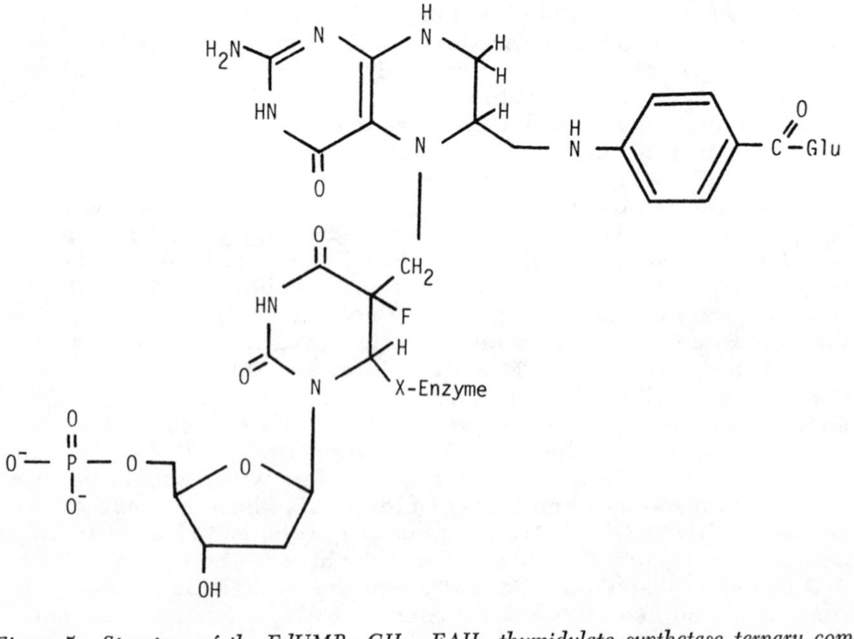

Figure 5. *Structure of the FdUMP · CH_2 · FAH_4 thymidylate synthetase ternary complex where X represents a nucleophile of one of the enzyme amino acids*

group of the cofactor. A similar structure was proposed from evidence obtained independently in another laboratory (20).

Referring to Figure 5, it is noted that the assigned structure for the E·FdUMP·CH$_2$FAH$_4$ complex is analogous to one of the proposed steady state intermediates of the normal enzymic reaction (viz II, Figure 2). They differ in that II possesses a proton at the 5-position of the nucleotide which is abstracted in a subsequent step, whereas the E·FdUMP·CH$_2$FAH$_4$ complex (Figure 5) possesses a stable fluorine at the corresponding position. Thus, it appears that FdUMP behaves as a "quasi-substrate" for this reaction. That is, it enters into the catalytic reaction as depicted for the substrate dUMP in Figure 2 up to the point where an intermediate is formed which can proceed no further; in effect, a complex is trapped which resembles a steady state intermediate (viz II, Figure 2) of the normal catalytic reaction.

We have recently obtained the fluorine-19 nmr spectrum of an FdUMP·CH$_2$FAH$_4$·peptide obtained upon proteolytic digestion of the FdUMP·CH$_2$FAH$_4$·thymidylate synthetase complex. As shown in Figure 6, the 94 MHz ^{19}F spectrum consists of a doublet of triplets located 87.2 ppm upfield of the external reference, trifluoroacetic acid. Our current interpretation of this ^{19}F spectrum is as follows: The doublet is caused by splitting of the ^{19}F resonance by the proton at the 6-position of the uracil ring (H$_A$) with a coupling constant J$_{AF}$ of 32.5 Hz. Each component of the doublet is split further into a triplet (intensity ratio (1:2:1) caused by coupling of the fluorine with the adjacent methylene protons (H$_B$) of the cofactor with the magnitude of the coupling constant H$_{BF}$ being 19.2 Hz. The innermost lines of the triplets overlap so the ^{19}F resonance appears to be a quintet with intensity ratio 1:2:2:2:1.

It has been well established that the trigonal geometry of the carbonyl atoms in uracil derivatives saturated across the 5,6-double bond results in a half-chair conformation with substituents on carbon atoms 5 and 6 staggered (28-30) as commonly found in cyclohexane.

The ^{19}F spectrum, in conjunction with previously reported ultraviolet and fluorescence spectral data (31), of the FdUMP· CH$_2$FAH$_4$·peptide yields definitive evidence for its structure. The doublet of triplets implies that the fluorine-bonded carbon is flanked by CH and CH$_2$ groups (i.e. CHCFCH$_2$). The CH, of course, occurs at the 6-position of the nucleotide which is attached to the nucleophile of the enzyme. The most logical assignment for the CH$_2$ is the bridging group between the nucleotide and cofactor as depicted for the native complex in Figure 5. Based on the stability of the FdUMP·CH$_2$FAH$_4$·peptide in the absence of antioxidants, we have surmised that the CH$_2$ group bridges the nucleotide to the 5-and not to the 10-nitrogen of the cofactor.

It is possible to assign the stereochemistry of the

^{19}F NMR

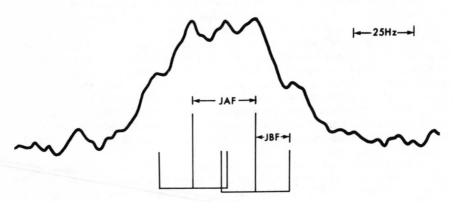

Figure 6. 94 MHz Fluorine-19 nmr spectrum of the FdUMP · CH₂FAH₄ · peptide

substituents which have added across the 5, 6-double bond of
FdUMP to give the ternary complex by comparing the observed
coupling constants with those from extensively studied models.
As shown in Figure 7, the 5-fluoro and 6-hydrogen of the
FdUMP·CH$_2$FAH$_4$·peptide are proposed to be in a <u>trans</u> pseudoaxial conformation. The enzyme nucleophile and cofactor are
therefore <u>trans</u> pseudoequatorial.

The addition of a nucleophile of thymidylate synthetase to
the 6-position of dUMP is a primary event in the enzyme-catalyzed reaction (Figure 8). The resultant carbanion (1) reacts
with CH$_2$ FAH$_4$ to produce an intermediate with a structure
analogous to that of the ternary complex formed with FdUMP
and the cofactor (2, 3). Abstraction of the 5-hydrogen followed
by a series of steps involving reduction of the one carbon unit
and elimination of the nucleophile results in the observed products of the reaction. These steps have been previously depicted in detail in Figure 2. With the logical assumption that
the normal enzyme-catalyzed reaction occurs in a manner similar to the formation of the thymidylate synthetase·FdUMP·
CH$_2$ FAH$_4$ complex, several details concerning the mechanism of
the normal enzymic reaction may be inferred.

First, the overall stereochemical pathway of the enzyme-
catalyzed reaction may be deduced. The addition of the nucleophile and cofactor across the 5, 6-double bond must occur in
a <u>trans</u> fashion. Consequently, the subsequent elimination of
the 5-hydrogen and the enzymic nucleophile must occur as a <u>cis</u>-
elimination .

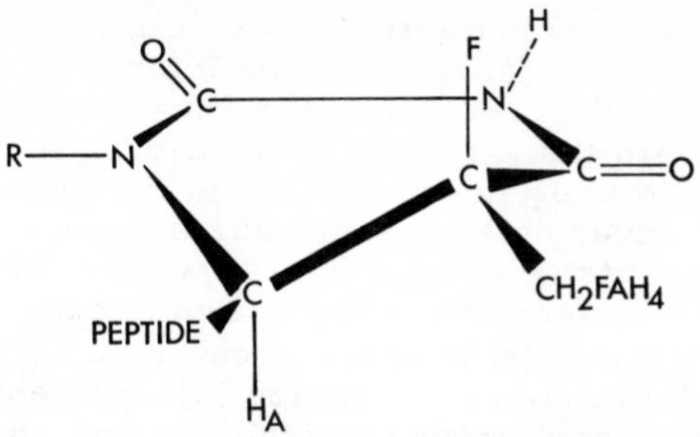

Figure 7. Stereochemical projection of the FdUMP · CH₂FAH₄ · peptide as determined by its ¹⁹F nmr spectrum; R = 5-phospho-2′deoxyribosyl

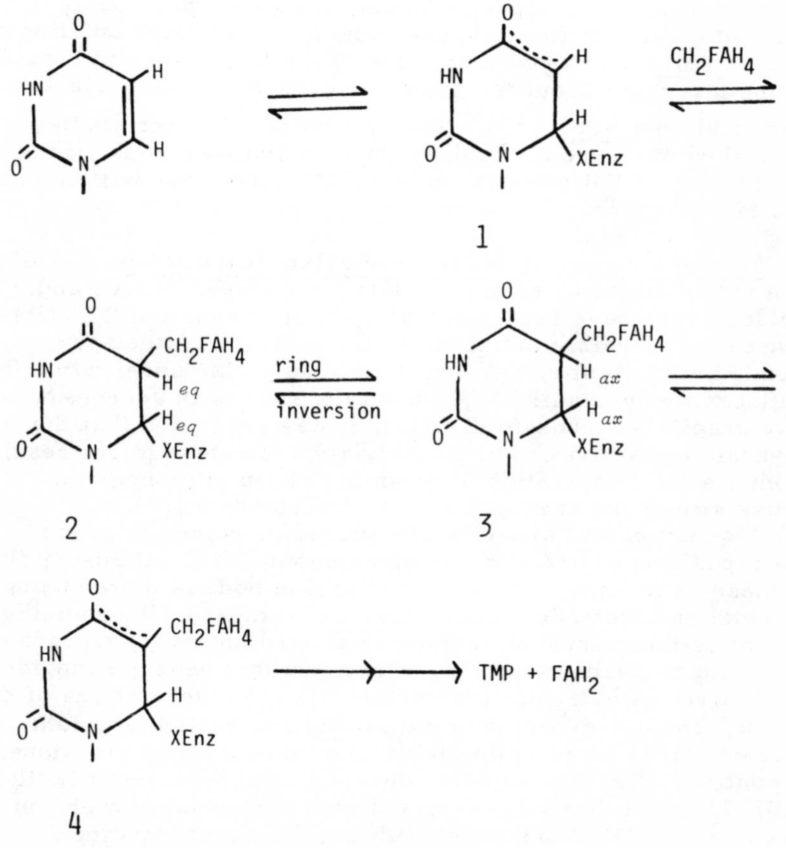

Figure 8

Second, the stereochemistry concerning the transient intermediates shown in Figure 8 may be inferred. The mechanistic details discussed below follow from the principle that a group reacting with the π-system of the uracil heterocycle approaches approximately perpendicular to the plane of the ring; by microscopic reversibility, a similar orientation is required when a group departs to reform the π-system. Thus, the initial attack of the nucleophile of the enzyme at the electrophilic 6-carbon of dUMP should be perpendicular to the plane of the heterocycle. The resultant carbanion (1) will be delocalized throughout the carbonyl groups and will be high in sp^2 character. The approach of $CH_2 FAH_4$ to the 5-position would be perpendicular to the plane of the ring and, based on the ^{19}F data presented above, <u>trans</u> to the nucleophile attached to the 6-position. As a result, as shown in structure 2, the cofactor would exist in a pseudoaxial position, and the 5-hydrogen would be pseudoequatorial. For the subsequent elimination reaction, the proton from the 5-position must be in the pseudoaxial position (3) prior to its abstraction to form the carbanion (4). This mechanistic interpretation necessitates a previously unrecognized conformational change, which occurs after addition of the cofactor but before abstraction of the 5-proton; and results in ring inversion about the 5-and 6-positions of the nucleotide intermediates (<u>i. e.</u> 2→3). The results of ^{19}F nmr studies described here are preliminary; a detailed nmr study of the interaction of FdUMP with thymidylate synthetase will be published elsewhere.

5-Trifluoromethyl-2'-deoxyuridylate (CF_3 dUMP).

CF_3 dUMP is a potent inhibitor of thymidylate synthetase. Reyes and Heidelberger have reported that upon preincubation CF_3 dUMP causes irreversible inhibition of thymidylate synthetase from Ehrlich Acsites cells (32). Based on the observation that trifluoromethyluracil (CF_3U) acylates amines in aqueous media to give uracil-5-carboxamides (33), it was suggested that the irreversible inactivation of thymidylate synthetase might result from a similar acylation of an amino group at or near the active site of the enzyme as shown in Figure 9 (32).

A question that arose is why the trifluoromethyl group at the 5-position of uracil derivatives should be at all susceptible to these reactions. The carbon-fluorine bond is quite strong (34) and an outstanding characteristic of most trifluoromethyl groups is their unusual resistance toward chemical degradation. As relevant examples, it is noted that benzotrifluorides, derivatives of 6-trifluoromethyluracils (35), derivatives of 2-trifluoromethyl-4-oxopyrimidines (36) and 5-trifluoromethyl-6-azauracil (37) are quite stable toward hydrolytic reactions. In contrast, CF_3U is rapidly converted into 5-carboxyuracil (CU) (33) in basic media and, although somewhat slower, nucleosides of CF_3U are converted into the corresponding

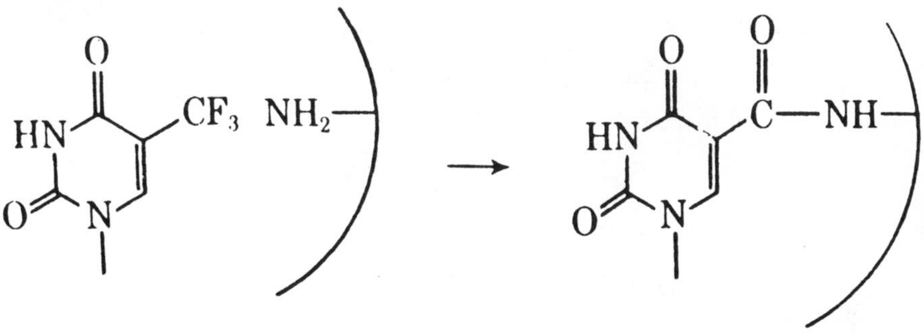

Figure 9

nucleosides of CU (38-40). In vivo, the metabolism of CF_3U and CF_3dUR provides CU and not the normal products expected from pyrimidines (41).

A number of other compounds have been reported to have reactive C-F groups (see reference 42). The reactivity of C-F bonds in most cases has been attributed to hyperconjugative effects (43, 44), hydrogen bonding effects (44, 45), and direct displacement (S_N2) reactions (46). Model reactions involving the hydrolysis of compounds possessing C-F bonds were examined in this labotatory in an attempt to understand the underlying features which resulted in the reactivity of some of these molecules (42). From such studies we proposed that C-F bond labilization usually involves one of several general mechanisms. (a) As depicted in Figure 10a, proton removal is at an atom α to the carbon bearing the fluorine atom with the resultant negative charge, either in a stepwise or concerted manner, aiding in the formation of an intermediate (fluoro) alkene. Depending on the stability of the alkene, it may or may not react with solvent. (b) The proton may be situated on an atom such that the negative charge resulting from the ionization of the proton can exert its influence through an extended π-system (Figure 10b). (c) When the compound is an allylic fluoride incapable of undergoing either of the mechanisms described above, it may undergo nucleophilic (Michael-type) attack at the β-carbon with assistance by the developing carbanion to give an intermediate similar to those previously described (Figure 10c). In any of the above, trifluoromethyl groups give carboxylic acids or derivatives, difluoromethyl groups give aldehydes or ketones, and fluoromethyl groups give alcohols or alkenes.

Most of the C-F bond cleavages thus far reported can be explained then in terms of the aforementioned mechanisms; the ability of the compounds to form olefinic intermediates of the type described appears necessary for such reactions to occur. The mechanism(s) by which the olefinic intermediates are transformed to products is believed to involve alternate addition of

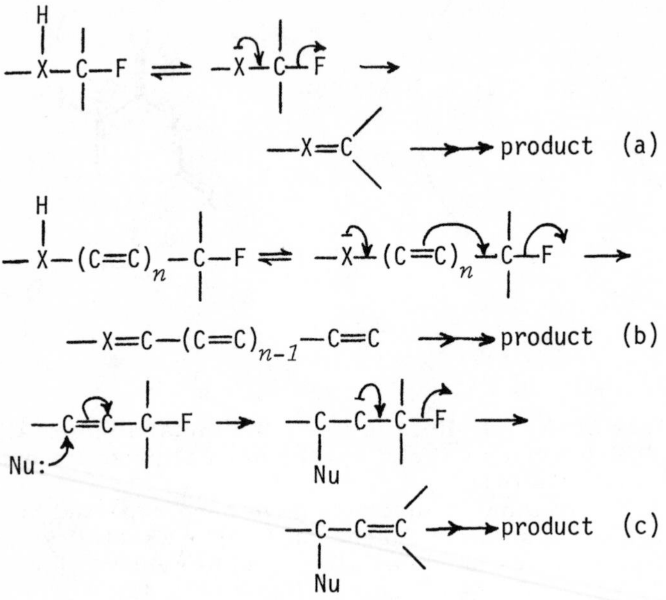

Figure 10

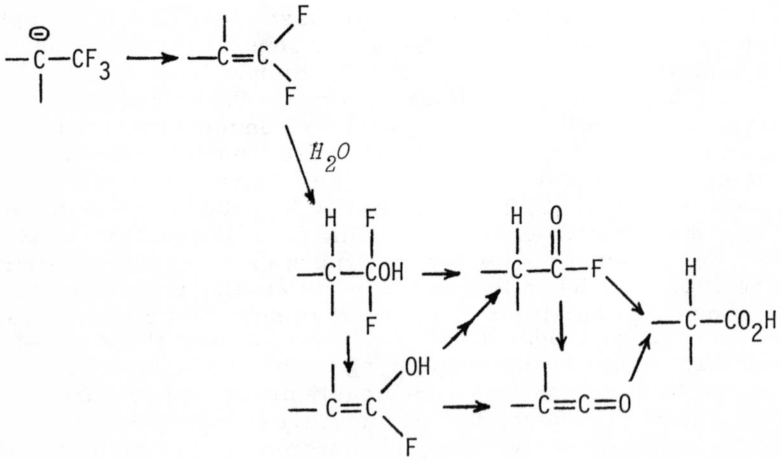

Figure 11

nucleophile (or solvent) to the intermediate, and elimination of fluoride ion. A possible mechanism for hydrolysis of the CF_3 group is depicted in Figure 11 and, as shown, may involve the intermediacy of acyl fluorides and ketenes in the transformation of a trifluoromethyl group to a carboxylate function, although these intermediates have, as yet, not been detected.

The above provided insight into the possible mechanisms by which CF_3dUMP might act as an acylating agent. To obtain direct supporting evidence, the mechanisms of hydrolytic reactions of 5-trifluoromethyluracil and its N-alkylated derivatives were examined in detail (47). The results of these studies are summarized below, and depicted in Figures 12-14.

All reactions appear to proceed by formation of a highly reactive intermediate having an exocyclic difluoromethylene group at the 5-position which subsequently reacts with water or hydroxide ion in a series of rapid steps to give corresponding 5-carboxyuracils. For those derivatives which possess an ionizable proton at the 1-position, the predominant mechanism involves ionization to the conjugate base and assistance by the 1-anion in the expulsion of fluoride ion (Figure 12). When ionization at the 1-position is precluded by the presence of an alkyl substituent (Figure 13), acylation reactions proceed by rate determining attack of hydroxide ion at the 6-position of the neutral or negatively charged (3-anion) heterocycle to provide the reactive intermediate. In order to obtain suitable intramolecular models, and to verify the primary site of reaction of 1-substituted derivatives, a series of 1-(ω-aminoalkyl)trifluoromethyluracils were prepared and their hydrolyses examined (Figure 14). Neighboring group participation was apparent where attack of the amino group on the 6-position of the heterocycle results in the formation of five-, six-, and seven-membered rings; in the case of 1-(3-aminopropyl)-5-trifluoromethyluracil, apparent first-order constants were more than 10^4 times greater than simple 1-alkyl derivatives not possessing a neighboring nucleophile.

With regard to the enzymic reaction, the salient feature of these studies is that the trifluoromethyl group of CF_3 dUMP derivatives would only behave as an acylating agent when a secondary driving force is furnished by reactions which occur at other parts of the heterocycle. That is, it is necessary that a nucleophile is added to the 6-position of the heterocycle; in this manner, the normally inert trifluoromethyl group would be converted into a highly reactive exocyclic difluoromethylene intermediate which might then acylate a nucleophilic group of the enzyme.

In accord with proposals for the involvement of nucleophilic catalysis in the enzymic reaction (viz I, Figure 2), these studies led us to propose a related minimal mechanism for the reported irreversible inactivation of thymidylate synthetase by CF_3dUMP. In the pathway depicted in Figure 15, it was suggested that juxtaposed within the active site, a nucleophilic group of the enzyme (:Z) adds to the 6-position of CF_3dUMP, promoting the expulsion of fluoride ion and the formation of a reactive exocyclic difluoromethylene intermediate similar to those encountered in our model studies. The reactive intermediate would then be trapped by a nucleophilic group of

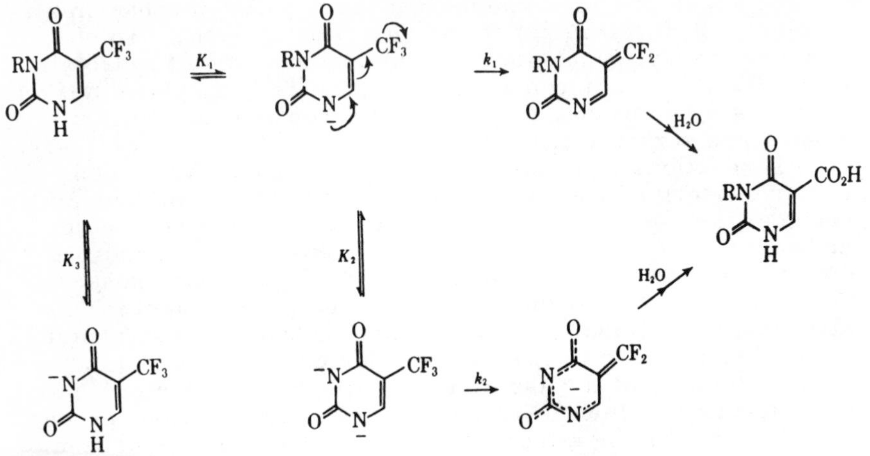

Figure 12

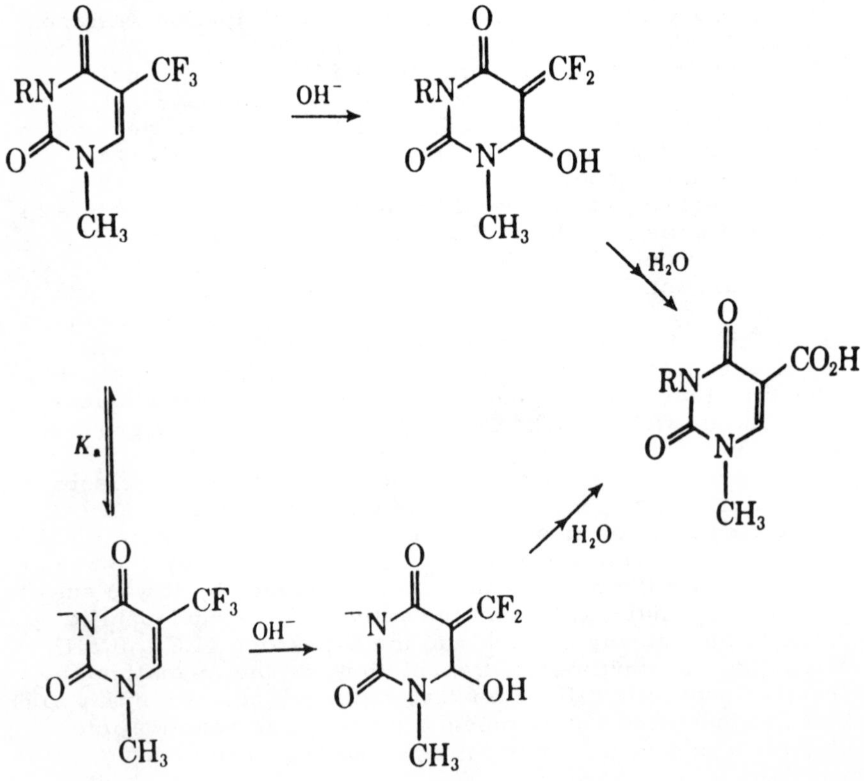

Figure 13

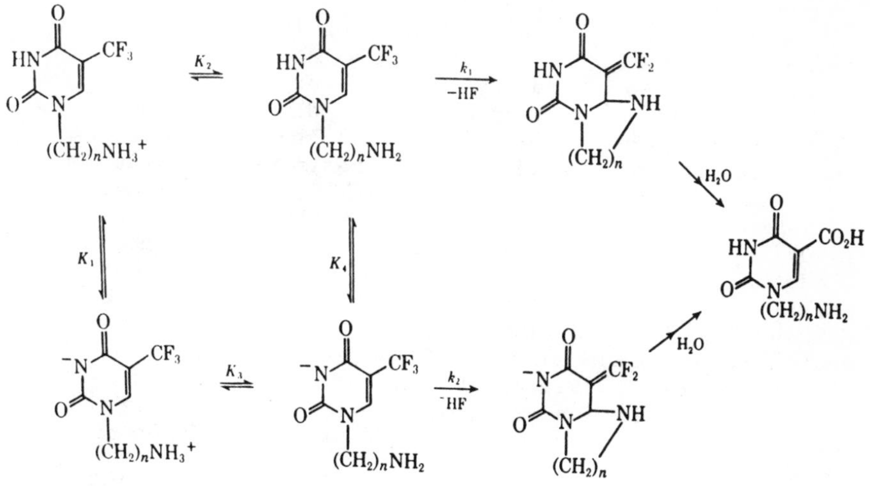

Figure 14

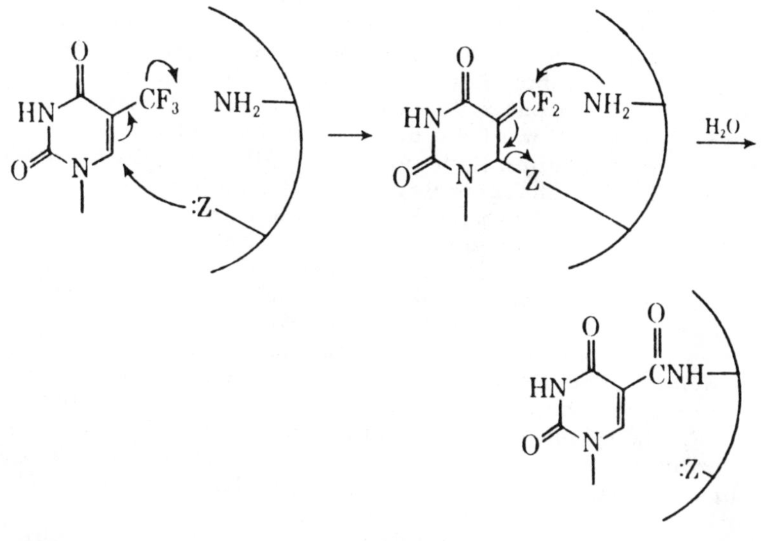

Figure 15

the enzyme to give, after a number of steps, the acylated enzyme.

Subsequent to completion of these model studies, the interaction of CF_3dUMP with thymidylate synthetase was reinvestigated in another laboratory using the enzyme from L. casei (21). These workers observed that CF_3dUMP, CH_2FAH_4 and thymidylate synthetase formed a tight ternary complex which was isolatable by disc gel electrophoresis under non-denaturing conditions. However, unlike the FdUMP·CH_2FAH_4·enzyme complex, no change in the difference spectra was observed when CF_3dUMP was used. Furthermore, gel electrophoresis in the presence of a protein denaturant resulted in apparent destruction of the complex. After denaturation of the complex, the nucleotide product was observed to migrate identically with authentic CF_3dUMP on DEAE-cellulose paper. From these results, it was concluded that C-F bonds of the nucleotide were not cleaved by the enzyme and a nucleophile of the enzyme did not add to the 6-position of CF_3dUMP.

Recent results obtained in this laboratory are not in accord with these findings. Although the mechanism of reaction of CF_3dUMP with thymidylate synthetase has not been ascertained at this time, it appears certain that C-F bond cleavage is catalyzed by the enzyme, probably via nucleophilic catalysis. Our preliminary results which lead to this conclusion are summarized below.

Contrary to the previous report (21), we observe that subtraction of the ultraviolet spectrum of enzyme and CH_2FAH_4 from that of the enzyme, CH_2FAH_4 and CF_3dUMP produces a difference spectra (Figure 16a) which is very similar to that observed with the FdUMP·CH_2FAH_4·enzyme ternary complex (Figure 4). As with the thymidylate synthetase·FdUMP·CH_2FAH_4 complex, there is the characteristic increase of absorbance at 330 nm and a decrease at 261 nm; the latter is in accord with saturation of the 5,6-double bond of the nucleotide. Upon addition of sodium dodecyl sulfate, the only differential absorbance is that characteristic of a nucleotide with $\lambda_{max} \simeq$ 267 nm; although this is upfield from the maximum of CF_3dUMP, we have not yet ascertained whether an alteration of structure is involved or whether the shift is artifactual.

In the absence of CH_2FAH_4, the incubation of thymidylate synthetase and CF_3dUMP (ratio 1:6) for 20 minutes at 22° results in 89% inactivation of the enzyme. This is in accord with previous reports on the effect of this nucleotide on the enzyme from Ehrlich ascites cells (32). The difference spectra of CF_3dUMP and enzyme vs enzyme is shown in Figure 16b. The maximum of CF_3dUMP at 261 nm decreases, and a transient broad peak appears which has absorbtion up to ca. 340 nm. After 1 hour, the final spectrum exhibits a maxima at 276 nm, resembling 5-acyl derivatives of dUMP. Paper

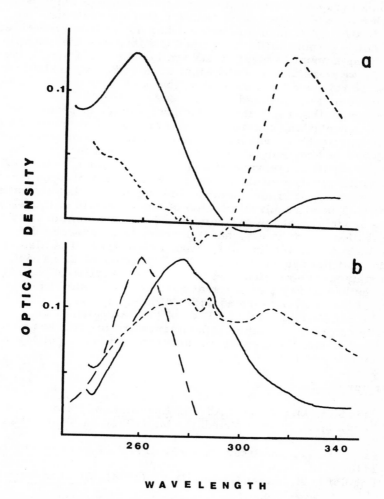

Figure 16. Ultraviolet difference spectra. (a) Dashed line: CF_3dUMP, CH_2FAH_4 and thymidylate synthetase vs. CH_2FAH_4 and thymidylate synthetase. Solid line: after treatment with sodium dodecyl sulfate. (b) Dashed line: CF_3dUMP and thymidylate synthetase vs. thymidylate synthetase after 20 seconds. Solid line: after 1 hr. Broken line: ultraviolet spectrum of CF_3dUMP.

chromatography of this reaction mixture shows a single spot which moves slightly slower than the starting material (CF_3dUMP). Although this product has not yet been identified, it is not 5-carboxy-dUMP. When CF_3dUMP was treated with a limiting amount of thymidylate synthetase (50:1) for 23 hours at 22°, we were able to detect that at least 0.3 equivalents of F^- were released, demonstrating that C-F bonds of the nucleotide were indeed labilized.

Although these results are too preliminary to permit definitive interpretation, certain conclusions may be reached and speculations may be forwarded. It is clear that the interaction of CF_3dUMP and thymidylate synthetase in the absence of cofactor may result in cleavage of C-F bonds of the nucleotide as well as inactivation of the enzyme. From the aforementioned model studies, it is most reasonable to propose that activation of the C-F bond requires addition of a nucleophile of the enzyme to the 6-position of the nucleotide. The mechanism of enzyme inactivation is not known, but it appears to be slower (10%) than the reaction leading to C-F bond cleavage. It is also apparent that the presence of CH_2FAH_4 modulates this reaction in some yet unknown manner; the enzyme·CF_3dUMP·CH_2FAH_4 complex has ultraviolet spectral qualities quite similar to those of the complex formed with FdUMP, indicating that the 5,6-double bond of CF_3dUMP is saturated in the ternary complex. Further experiments are in progress which aim to elucidate the mechanism of interaction of CF_3dUMP with thymidylate synthetase.

Literature Cited

1. Friedkin, M., (1973), Advan. Enzymol. 38, 235.
2. Pastore, E. J., and Friedkin, M. (1962), J. Biol. Chem. 237, 3802.
3. Santi, D. V., and Brewer, C. F. (1973), Biochemistry 12, 2416.
4. Pogolotti, A. L., and Santi, D. V. (1974), Biochemistry 13, 456.
5. Santi, D. V., and Brewer, C. F. (1968), J. Amer. Chem. Soc. 90, 6236.
6. Kalman, T. I. (1971), Biochemistry 10, 2567.
7. Benkovic, S. J., and Bullard, W. P. (1973), Prog. Bioorg. Chem. 2, 134.
8. Crusberg, T. C., Leary, R., and Kisliuk, R. L. (1970), J. Biol. Chem. 245, 5292.
9. Dunlap, R. B., Harding, N. G. L., and Huennekens, F. M. (1971), Biochemistry 10, 88.
10. Leary, R. P., and Kisliuk, R. L. (1971), Prep. Biochem. 1, 47.
11. Galivan, J. H., Maley, G. F., and Maley, F. (1975), Biochemistry 14, 3338.

12. Cohen, S.S. , Flaks, J. G. , Barner, H.D. , Loeb, M.R. , and Lichenstein, J. (1958), Proc. Nat. Acad. Sci. U.S. 44, 1004.
13. Heidelberger, C. , Kaldor, G. , Mukherjee, K.L. , and Danenberg, P.B. (1960), Cancer Res. 20, 903.
14. Blakley, R.L. (1969). The Biochemistry of Folic Acid and Related Pteridines, New York, N.Y. , American Elsevier.
15. Fox, J. J. , Miller, N.C. , and Cushley, R.J. (1966), Tetrahedron Lett. , 4927
16. Otter, B.A. , Falco, E.A. , and Fox, J.J. (1969), J. Org. Chem. 34, 1390.
17. Reist, E.J. , Benitez, A. , and Goodman, L. (1964), J. Org. Chem. 29, 554.
18. Santi, D.V. , and McHenry, C.S. (1972), Proc. Nat. Acad. Sci. U.S. 69, 1855.
19. Santi, D.V. , McHenry, C.S. , and Sommer, H. (1974), Biochemistry 13, 471.
20. Langenbach, R.J. , Danenberg, P.V. , and Heidelberger, C. (1972), Biochem. Biophys. Res. Commun. 48, 1565.
21. Danenberg, P.V. , Langenbach, R.J. , and Heidelberger, C. (1974), Biochemistry 13, 926.
22. Rose, I.A. , O'Connell, E.L. , Litwin, S. , and Bar Tana, J. (1974), J. Biol. Chem. 249, 5163.
23. Santi, D.V. , McHenry, C.S. , and Perriard, E.R. (1974), Biochemistry 13, 467.
24. Aull, J.L. , Lyon, J.A. , and Dunlap, R.B. (1973), Biochem. Jour. 19, 210.
25. Sharma, R.K. , and Kisliuk, R.L. (1973), Fed. Proc. , Fed. Amer. Soc. Exp. Biol. 31, 591.
26. Aull, J.L. , Lyon, J.A. , and Dunlap, R.B. (1974), Arch. Biochem. Biophys. 165, 805.
27. Loeble, R.B. , and Dunlap, R.B. (1972), Biochem. Biophys. Res. Commun. 49, 1671.
28. Rouillier, P. , Delmau, J. , and Nofre, C. (1966), Bull. Soc. Chim. (France), 3515.
29. Furburg, S. , and Janson, L.H. (1968), J. Amer. Chem. Soc. 90, 470.
30. Katritzky, A.R. , Nesbit, M.R. , Kurtev, B.J. , Lyapova, M. , and Pojarlieff, I.G. (1969), Tetrahedron 25, 3807.
31. Wahba, A.J. , and Friedkin, M. (1961), J. Biol. Chem. 236, PC 11.
32. Reyes, P. , and Heidelberger, C. (1965), Mol. Pharmacol. 1, 14.
33. Heidelberger, C. , Parsons, D.G. , and Remy, D.C. (1964), J. Med. Chem. 7, 1.
34. Pauling, L. (1960), The Nature of the Chemical Bond, Ithaca, N.Y. , Cornell University Press, 85.
35. Giner-Sorolla, A. , and Bendich, A. (1958), J. Amer. Chem. Soc. 80, 5744.
36. Barone, J.A. (1963), J. Med. Chem. 6, 39.

37. Dipple, A., and Heidelberger, C. (1966), J. Med. Chem. 9, 715.
38. Shen, T. Y., Ruyle, W. V., and Lewis, H. M. (1965), J. Org. Chem. 30, 835.
39. Khwaja, T. A., and Heidelberger, C. (1969), J. Med. Chem. 12, 543.
40. Ryan, K. J., Acton, E. M., and Goodman, L. (1966), J. Org. Chem. 31, 1181.
41. Heidelberger, C., Boohar, J., and Kampschroer, D. (1965), Cancer Res. 25, 377.
42. Sakai, T. T., and Santi, D. V. (1973), J. Med. Chem. 16, 1079.
43. Roberts, J. D., Webb, R. L., and McElhill, E. A. (1950), J. Amer. Chem. Soc. 72, 408.
44. Filler, R., and Novar, H. (1960), Chem. Ind. (London), 1273.
45. Jones, R. (1947), J. Amer. Chem. Soc. 69, 2347.
46. Nestler, H. J., and Garrett, E. R. (1968), J. Pharm. Sci. 57, 1117.
47. Santi, D. V., and Sakai, T. T. (1971), Biochemistry 10, 3598.

The Effect of Aliphatic Fluorine on Amine Drugs

RAY W. FULLER and BRYAN B. MOLLOY

The Lilly Research Laboratories, Eli Lilly & Co., Indianapolis, Ind. 46206

Fluorine is the most electronegative element (1) (Table I), and its presence in a molecule can greatly affect the ionization of acids and bases (2). The withdrawal of electrons toward fluorine on a carbon atom adjacent to a carboxyl carbon or an amino-bearing carbon increases the carboxyl group's ability to release a proton and decreases the amine's ability to accept a proton. The carboxylic acid becomes a stronger acid, and the amine becomes a weaker base. For example, the pK_a of ethylamine is >10 compared to 5.7 for β,β,β-trifluoroethylamine (3).

We have studied some sympathomimetic amine drugs having reduced basicity due to fluorine substituted on the β carbon. The influence of ionization on the activity of sympathomimetic amines has previously been considered (4), but with the exception of compounds having substituents directly on the nitrogen, no such amines having pK_a values below physiological pH appear to have been available (5). An advantage of the β,β-difluoro compounds is that the substitution is in a position of the molecule that is not a major site of metabolic attack nor a site known to be involved in binding to physiological receptors, and the small size of the fluorine atoms (Table I) results in minimal steric alteration in the molecule. Thus the changes in the pharmacological properties of the amines as a result of β,β-difluoro substitution probably are due to the change in ionization.

In solution an amine exists in the following equilibrium,

$$RNH_3+ \underset{+H+}{\overset{-H+}{\rightleftharpoons}} RNH_2$$

the position of the equilibrium depending on the pK_a
of the amine and the pH of the solution. When the pH
is equal to the pK_a, 50% of the drug molecules are protonated (cationic) and 50% are nonprotonated (neutral).
When the pH is one unit below the pK_a, just over 90%
of the molecules exist in the cationic form, and when
the pH is two units below the pK_a more than 99% of the
molecules are charged. In this paper we compare some
primary amines with pK_a values well above physiological
pH to their β,β-difluoro derivatives having pK_a values
below physiological pH. The parent amines are nearly
completely cationic whereas the β,β-difluoro derivatives exist predominantly as neutral molecules at
physiological pH. Presumably as a result of this
charge difference, the pharmacologic properties of the
drugs--both the biological fate and the biological
actions of the drugs--are dramatically altered.

TABLE I

Fluorine in Relation to Other Halogen Atoms[1]

Atom	Van der Waals radius, $\overset{o}{A}$	Electronegativity value
F	1.35	4.0
Cl	1.80	3.0
Br	1.95	2.8
I	2.15	2.5
H	1.2	2.1

[1]Data from (1).

Amphetamine

The effect on ionization of amphetamine produced
by the presence of one or two fluorine atoms on the
β carbon is shown in Table II. A single fluorine atom
reduces the pK_a value by more than one pH unit. The
pK_a of the monofluoro derivative of amphetamine is
still above physiological pH, however, so that most of

the monofluoro compound is ionized at physiological pH.
More than 99% of the molecules of amphetamine exist as
cations at physiological pH. In contrast, the difluoro
compound has a pKa below physiological pH and thus
would exist mainly as a neutral molecule rather than as
a cation in the body.

TABLE II

*Effect of Fluorine Substution on the
Ionization of Amphetamine*

Compound	pK_a	Distribution of Ionic Forms at pH 7.4	
		RNH_2	RNH_3+
benzene–CH_2CHNH_2 | CH_3	9.45	1	99
benzene–$CHCHNH$ | | F CH_3	8.35	10	90
benzene–CF_2CHNH_2 | CH_3	6.97	73	27

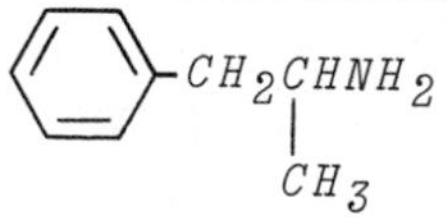

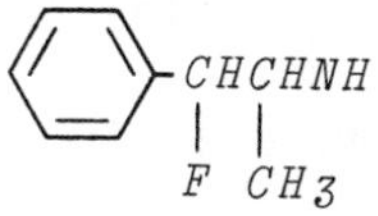

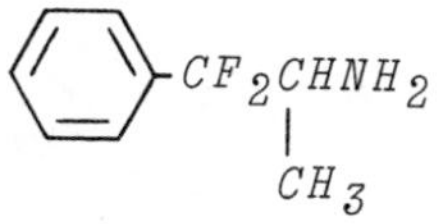

The interaction of amphetamine with biological
macromolecules ought to be profoundly affected by the
fluorine substitution and attendant ionic changes,
since the transport and enzymic systems that determine
distribution and metabolism of the drug should act
differently on a neutral unprotonated amine compared
to a cation. A number of studies on the interaction
of amphetamine and its fluorinated derivatives with
enzymes and other proteins in vitro and of the pharmacologic characteristics of these drugs in vivo have
borne out the expected alterations in properties of
amphetamine resulting from fluorine substitution.

In Vitro Interactions. β,β-Difluoro substitution
increases the activity of amphetamine as an in vitro
substrate for lung N-methyltransferase (6) or liver
microsomal deaminase (7). On the other hand, β,β-

difluoro substitution decreases the activity of amphetamine as an inhibitor of mitochondrial monoamine oxidase (Table III). Likewise, β,β-difluoro substitution
diminishes the binding of amphetamine to bovine serum
albumin (Table III). Thus reducing the pKa of amphetamine by β,β-difluoro substitution can either enhance
or inhibit its interactions with various biological
macromolecules in vitro.

TABLE III

Effect of β,β-Difluoro Substitution on Two Properties
of Amphetamine: Monoamine Oxidase Inhibition
and Binding to Bovine Serum Albumin

Drug	MAO Inhibition pI50	% Bound to Albumin
Amphetamine	3.23	82
β,β-Difluoroamphetamine	2.56	45

MAO inhibition was measured with rat liver mitochondrial enzyme and ^{14}C-tryptamine as substrate. The
pI50 value is the negative log of the molar concentration of inhibitor producing 50% inhibition. Binding to
4% bovine serum albumin in pH 7.4 sodium phosphate buffer was measured with 10 micromolar drug concentration;
after filtration through Aminco Centriflo ultrafiltra-
tion cones drug levels were assayed colorimetrically
with methyl orange.

    In Vivo Properties. The effect of β-fluorine sub-
stitution on the distribution of amphetamine among
various body tissues after drug administration to experimental animals was striking. Figure 1 shows the
tissue distribution of amphetamine, β-fluoroamphetamine,
and β,β-difluoroamphetamine in rats one hour after the
drugs were injected intraperitoneally at equimolar
doses. Highest tissue levels of amphetamine were in
lung, and lowest levels were in the fat, all tissues
having higher levels than blood. The relative distribution of the monofluoro derivative was similar, levels
in all tissues except lung being about the same as
those of amphetamine. This similarity is expected
since the monofluoro compound, like amphetamine, is
mostly protonated at physiological pH. In marked contrast was the distribution of the difluoro derivative.

Highest levels of this drug were in fat, the tissue
that contained lowest levels of the other two amines.
Levels of the difluoro compound in other tissues were
consequently reduced; for example, the levels in brain
were only about one-fourth those of amphetamine. This
distribution at one hour is similar to that seen at
other times (7), i.e. there is little redistribution
of drug and the rate of disappearance of the drugs from
all tissues is about the same.

Since the predominant pharmacologic effects of
amphetamine result from its action on the brain, we
compared regional distribution of amphetamine and β,β-
difluoroamphetamine in brain (Table IV). The drugs
were given at doses chosen to produce comparable whole
brain levels. Some differences in distribution among
the various anatomic regions were noted, but these were
not as great as the differences in Figure 1 among
various organs. The subcellular distribution of
amphetamine in brain was not significantly altered
by difluoro substitution (7).

TABLE IV

Regional Distribution of Amphetamine and
β,β-Difluoroamphetamine in Rat Brain

Brain Region (% of Total Brain Weight)	Drug Level, nanomoles/g	
	Amphetamine	*β,β-Difluoro-amphetamine*
Cerebral hemispheres (62-67%)	46 ± 4	57 ± 2
Cerebellum (14-17%)	18 ± 6	33 ± 3
Midbrain (8-10%)	54 ± 4	92 ± 6
Brain stem (6-10%)	65 ± 20	71 ± 7
Hypothalamus (3-4%)	86 ± 11	84 ± 6

Drugs were injected i.p. (amphetamine at 10 mg/kg,
β,β-difluoroamphetamine at 40 mg/kg) 1 hour before the
rats were killed. Mean values and standard errors for
5 rats per group are shown.

Just as the tissue distribution of amphetamine is changed by β,β-difluoro substitution, so also is the metabolism of the drug changed. Although the half-lives of amphetamine and β,β-difluoroamphetamine are about the same in rats, the pathways of metabolism for the two drugs are different (Table V). Amphetamine is metabolized in the rat predominantly by hydroxylation on the para position of the aromatic ring, whereas β,β-difluoroamphetamine appears not to be metabolized by para-hydroxylation at all (7). Instead the difluoro compound is metabolized rapidly by oxidative deamination (7). Tissue levels of amphetamine and its half-life are increased by agents like desmethylimipramine that inhibit aromatic hydroxylation, whereas tissue levels of β,β-difluoroamphetamine are increased by typical inhibitors of microsomal enzymes like SKF 525A and DPEA (7). Induction of liver microsomal enzymes by chronic phenobarbital treatment markedly increases the rate of removal of β,β-difluoroamphetamine from tissues after it is injected into rats but has no effect on the rate of removal of amphetamine (8). β,β-Difluoro substitution increases the activity of amphetamine as a substrate for microsomal deaminases, an effect that can be shown in vitro. Apparently the difluoro substitution abolishes the activity of amphetamine as a substrate for the aromatic hydroxylating system, though this system is difficult to study in vitro. The failure of the difluoro compound to be hydroxylated in vivo may be due to one or both of the following reasons: (a) it is more readily deaminated than is amphetamine, (b) it may be less readily hydroxylated than amphetamine.

TABLE V

Effect of β,β-Difluoro Substitution on Amphetamine Metabolism in the Rat

Drug	Half-life in Brain	Major Metabolic Route
Amphetamine	1.3 hrs	Ring hydroxylation
β,β-Difluoro-amphetamine	1.5 hrs	Oxidative deamination

β,β-Difluoroamphetamine has been shown to increase locomotor activity in mice to the same extent as amphetamine (9), though approximately four times higher doses of the difluoro compound have to be given. Likewise about four times higher doses of the difluoro compound are required to produce equivalent drug levels in brain compared to amphetamine, so the intrinsic ability of the difluoro compound to cause CNS stimulation seems about equal to that of amphetamine.

In mice, the half-life of the difluoro compound (0.3 hours) is much shorter than is that of amphetamine (0.9 hours). Apparently this difference is related to the fact that mice have the enzymic machinery to metabolize amphetamine by oxidative deamination (10), and since the difluoro derivative is a much better substrate for deamination it is metabolized much more readily in mice than in rats. Otherwise the comparison of amphetamine and β,β-difluoroamphetamine yields similar results in mice and in rats.

One use that has been made of β,β-difluoroamphetamine is as a tool to elucidate interrelationships among the various actions of amphetamine. Gessa, Clay and Brodie (11) had suggested that the hyperthermia following amphetamine injection into rats was a direct consequence of the elevation of plasma free fatty acids. We found that β,β-difluoroamphetamine injected at an equimolar dose elevated free fatty acids to the same extent as amphetamine but produced no hyperthermia (12), showing that these two effects of amphetamine were completely dissociable by virtue of difluoro substitution. Presumably the hyperthermic response to amphetamine is due to an action in brain, hence β,β-difluoroamphetamine produces hyperthermia only when it is injected at higher doses.

Another difference between amphetamine and β,β-difluoroamphetamine is the inability of the latter drug to cause depletion of norepinephrine levels in brain and heart (Table VI). High doses of amphetamine have long been known to deplete norepinephrine in these tissues (13). We gave the difluoro compound at four times the dose of amphetamine to produce equivalent drug levels in brain and heart but still found no reduction of norepinephrine levels. The precise mechanism(s) by which amphetamine lowers norepinephrine levels is still uncertain, but hydroxylated metabolites may play a role in this effect. The failure of β,β-difluoroamphetamine to lower norepinephrine could then be explained because it is not metabolized by hydroxylation.

TABLE VI

Inability of β,β-Difluoroamphetamine to Cause
Amphetamine-Like Depletion of Tissue Norepinephrine

Treatment Group	Norepinephrine Levels, $\mu g/g$	
	Heart	Brain
Control	$0.89\pm.05$	$0.40\pm.01$
dl-Amphetamine 0.1 mmole/kg	$0.58\pm.01$ $(P<.001)$	$0.35\pm.01$ $(P<.05)$
dl-β,β-Difluoro-amphetamine 0.4 mmole/kg	$0.91\pm.06$ (ns)	$0.41\pm.03$ (ns)

Drugs were injected i.p. 6 hours before the rats were killed. Mean values and standard errors for 5 rats per group are shown.

Phenethylamine

The ionization of phenethylamine is affected by β-fluorine substitution in much the same way as that of amphetamine (Table VII). A single fluorine reduces the pK_a, and a second fluorine further reduces the pK_a to below physiological pH.

TABLE VII

Effect of Fluorine Substitution on the
Ionization of Phenethylamine

Compound	pK_a	Distribution of Ionic Forms at pH 7.4	
		RNH_2	RNH_3+
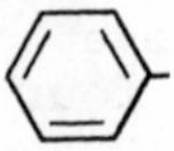 ⬡-$CH_2CH_2NH_2$	9.55	1	99
⬡-$CHCH_2NH_2$ $\|$ F	8.20	13	87
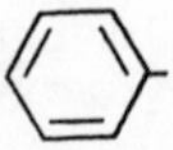 ⬡-$CF_2CH_2NH_2$	6.75	82	18

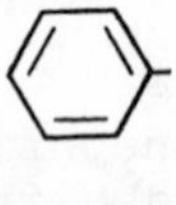

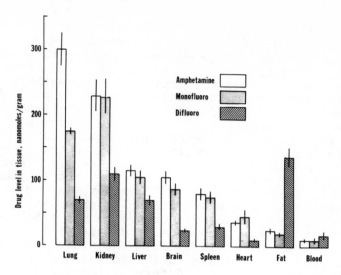

Figure 1.   Tissue distribution of amphetamine in rats as affected by
fluorine substitution on the β carbon.
Drugs were injected i.p. at 0.1 mmole/kg 1 hr before groups of 5
rats were killed. Mean values and standard errors are shown (7).

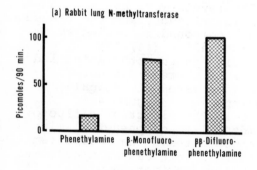

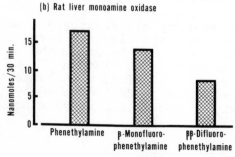

Figure 2.   Effect of β fluorine atoms on the
activity of phenethylamine as an enzyme
substrate.
The phenethylamines were compared at 4
mM with the N-methyltransferase and at
1 mM with monoamine oxidase.

In Vitro Interactions. The effect of β-fluorine substitution on the activity of phenethylamine as a substrate for two enzyme systems in vitro is shown in Figure 2. Addition of one and two fluorine atoms progressively reduced the activity of phenethylamine as a substrate for mitochondrial monoamine oxidase but progressively increased activity as a substrate for lung N-methyltransferase. The latter enzyme probably is not important in phenethylamine metabolism in vivo but illustrates that lowering the pKa of the amine can increase as well as decrease its affinity for enzymes. The ability of monofluoro and difluoro phenethylamines to inhibit the oxidation of ^{14}C-phenethylamine by three preparations of monoamine oxidase is shown in Figure 3. There was a substantial difference in the inhibitory activity of the two compounds, and this difference appeared to result entirely from the difference in pKa values. Plotting percent inhibition as a function of concentration of protonated form of the two inhibitors revealed that the points fell along the same line; this finding suggests that the RNH3+ form is the active inhibitor of monoamine oxidase and that this form of both compounds has equal affinity for the enzyme.

In Vivo Properties. Phenethylamine itself was very rapidly degraded by monoamine oxidase when it was injected into animals, but the difluoro derivative was less readily destroyed. Its levels were higher than those of phenethylamine in all tissues, and the tissue distribution of the difluoro compound resembled very much that of β,β-difluoroamphetamine (14).

Thus difluoro substitution on phenethylamine led to greater biological stability whereas difluoro substitution on amphetamine decreased biological stability at least in some species. Figure 4 illustrates this difference in the effect of difluoro substitution on in vitro metabolism of phenethylamine and amphetamine by liver homogenates. The metabolism being measured is deamination in both cases; microsomal enzymes are responsible for the deamination of amphetamine, whereas mitochondrial monoamine oxidase is responsible for the deamination of phenethylamine. β,β-Difluoroamphetamine was more rapidly destroyed than amphetamine, but β,β-difluorophenethylamine was less rapidly destroyed than phenethylamine. In rats, the additional metabolic route of para-hydroxylation operates on amphetamine in vivo, so in fact amphetamine is destroyed at about the same rate as difluoroamphetamine in rats but more slowly in mice. Phenethylamine

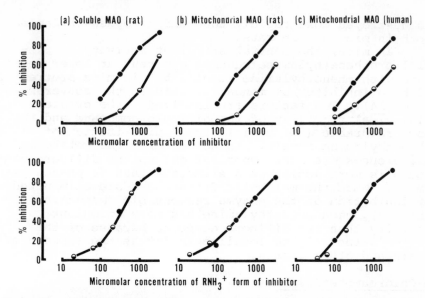

Figure 3   *Inhibition of three monoamine oxidase preparations by β-monofluoro-phenethylamine (dots) and by β,β-difluoro-phenethylamine (half circles). The substrate was phenethylamine (.2 mM). Total inhibitor concentration is shown at the top, and the concentration of the protonated form of the inhibitor (pH 7.4) is shown at the bottom. Monoamine oxidase was solubilized from rat liver mitochondria (a), or intact liver mitochondria were used as enzyme source (b) and (c).*

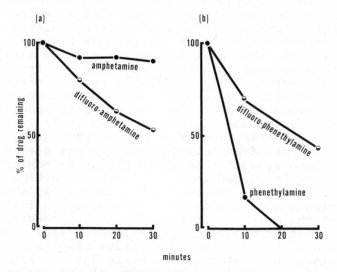

Figure 4.   *Effect of β,β-difluoro substitution on the* in vitro *degradation of (a) amphetamine and (b) phenethylamine by rat liver homogenates.*
The amines were added at .125 mM concentrations to liver homogenates.

is destroyed more rapidly than difluorophenethylamine
both in mice and in rats.
     In mice, the central stimulant activity of β,β-
difluorophenethylamine becomes apparent at lower doses
than with phenethylamine itself, but in mice pretreated
with an inhibitor of monoamine oxidase the converse is
true (14). Two factors are involved: the greater
biological stability of the difluoro compound and the
more favorable (for brain) tissue distribution of
phenethylamine itself. In normal mice the metabolic
differences are more important and so the difluoro compound is more active as a stimulant than is phenethylamine. When the metabolic differences are eliminated
by inhibition of the enzyme responsible (monoamine
oxidase), then phenethylamine has more stimulant
activity than the difluoro compound because of the
latter's tendency to localize in fat at the expense of
brain and other organs.

p-Chloroamphetamine

     p-Chloroamphetamine is a drug of special interest
because of its selective effects on serotonin neurons
in brain. p-Chloroamphetamine leads to a rapid
decrease in the levels of serotonin and its major
metabolite, 5-hydroxyindoleacetic acid, in brain. In
the rat and some other species, p-chloroamphetamine
leads further to cytotoxic destruction of serotonin
neurons as indicated by remarkably long-lasting
decreases in parameters specifically associated with
serotonin neurons in brain (tryptophan hydroxylase,
serotonin and 5-hydroxyindoleacetic acid levels; high-
affinity serotonin uptake) (15,16) and by direct
histologic evidence (17). Thus it was of interest to
study the difluoro derivative of p-chloroamphetamine.
     The reduction of the pK_a of p-chloroamphetamine
by β,β-difluoro substitution is shown in Table VIII.
The tissue distribution of p-chloroamphetamine, which
resembles that of amphetamine itself, was altered by
the difluoro substitution. β,β-Difluoro-p-chloro-
amphetamine localized to an even greater extent in fat
than did β,β-difluoroamphetamine, apparently through
a contribution of the chlorine to the overall lipophilicity of the molecule in addition to the reduced ionization due to the difluoro substitution (18). The
half-life of the difluoro derivative was less than that
of p-chloroamphetamine in rat brain (18), presumably
because of more facile deamination of the difluoro
compound. When high doses of the difluoro compound
were injected to produce drug levels in brain

TABLE VIII

Effect of β β-Difluoro Substitution on the
Ionization of p-Chloroamphetamine

Compound	pK_a	Distribution of Ionic Forms at pH 7.4	
		RNH_2	RNH_3+
p-Chloroamphetamine	9.3	1	99
β,β-Difluoro-p-Chloroamphetamine	6.8	80	20

equivalent to those of p-chloroamphetamine (18) sero-tonin and 5-hydroxyindoleacetic acid levels in brain were depleted to about the same extent as by p-chloro-amphetamine at 6 hours (Table IX). However, there was

TABLE IX

Brain Serotonin and 5-Hydroxyindoleacetic Acid
Levels after Injection of p-Chloroamphetamine
or its β,β-Difluoro Derivative

Drug Treatment	Serotonin	5HIAA
	% of Control	
p-Chloroamphetamine (0.1 mmole/kg)		
6 hrs	46 ± 3*	54 + 2*
24 hrs	43 ± 3*	50 ± 2*
β,β-Difluoro-p-chloroamphetamine (0.4 mmole/kg)		
6 hrs	44 ± 1*	56 ± 3*
24 hrs	97 ± 6	95 ± 2

*Significant drug effect, P<.01 (5 rats/group).

a difference in duration of the action of the two
compounds. With the difluoro derivative, brain
5-hydroxyindole levels had returned to normal within
24 hours (Table IX). On the other hand, p-chloroamphetamine lowers brain 5-hydroxyindoles not only 24
hours (Table IX) but for several months (15,16) after
a single injection. This long-lasting effect appears
to be due to a toxic action of p-chloroamphetamine on
serotonin neurons (17). Thus the difluoro compound has
the same short-term effect as p-chloroamphetamine on
brain serotonin but lacks the neurotoxic effect of the
parent drug. Whether this difference is due to the
inability of the difluoro compound to be converted to a
neurotoxic metabolite as happens with p-chloroamphetamine or to some other factors cannot be known at
present.

The initial lowering of brain serotonin by both
p-chloroamphetamine and β,β-difluoro-p-chloroamphet-
amine depends on the active transport of those drugs
into serotonin neurons, since their effects are blocked
by inhibition of that active uptake (19). Continual
reuptake of p-chloroamphetamine is necessary for the
depletion of serotonin to be maintained, since treatment with an uptake inhibitor at early times after
p-chloroamphetamine injection can reverse the depletion
of serotonin (20). Thus one possibility is that the
difluoro compound is not so efficiently re-taken up by
the membrane pump on the serotonin neuron. Consistent
with this possibility is the recent finding (D. T. Wong
and F. P. Bymaster, unpublished data) that the difluoro
compound has only about one-tenth of the affinity for
the serotonin uptake pump as does p-chloroamphetamine
(determined by their relative ability to inhibit high-
affinity serotonin uptake into synaptosomes in vitro).

Other Amines

We have been interested in applying the pK_a lowering through β,β-difluoro substitution to amine drugs
whose localization in adipose tissue would be expected
to enhance their pharmacological actions, i.e. whose
target tissue would be fat.

One line of investigation has dealt with agents
that inhibit lipolysis. Nicotinic acid (I) lowers
serum free fatty acid levels and is used in the treatment of hyperlipoproteinemia (21). One of the side
effects that limits the usefulness of nicotinic acid is
flushing due to vasodilatation (21). We knew that
3-aminomethyl-pyridine (II) had antilipolytic activity
like nicotinic acid both in vitro and in vivo; the

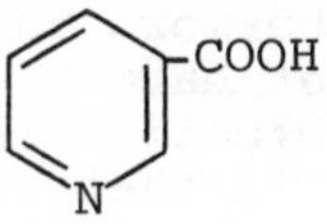

I

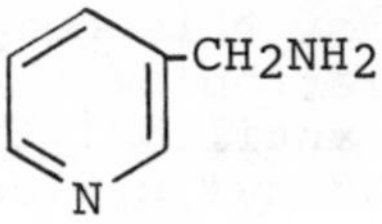

II

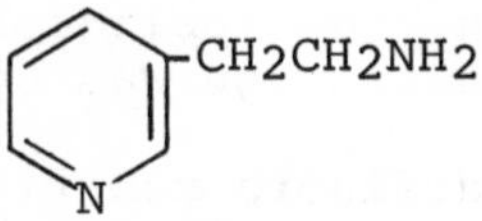

III

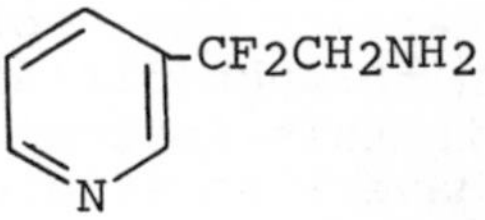

IV

amine is metabolized rapidly and completely to nicotinic acid in vivo, so its in vivo activity is due
almostly entirely to nicotinic acid. This amine then
offers no advantages over nicotinic acid as an antilipolytic drug.

We thought that substitution of fluorines to
reduce the pK_a of an amine of this sort might achieve
two objectives: (1) retard the oxidative deamination
of the amine and (2) cause the amine to localize pre-
ferentially in adipose tissue. The ultimate purpose
would be to reduce side effects while retaining the
antilipolytic activity of nicotinic acid. Though the
precise mechanisms of flushing caused by nicotinic acid
are not known, it seemed likely that an amine rather
than an acid--and in particular an amine that was concentrated mostly in fat--might be free of this side
effect.

To approach this objective, it was first necessary
to lengthen the side chain of 3-aminomethylpyridine to
provide a site for fluorine substitution. 3-Amino-
ethyl-pyridine (III) was synthesized and found to have
activity as an antilipolytic agent in vivo and in
vitro; the activity in vitro indicates that the amine
itself is active and does not require conversion to the
acid. Finally the β,β-difluoro derivative of 3-amino-
ethyl-pyridine (IV) was synthesized and studied as an
antilipolytic drug. Some methodologic difficulties
were encountered in measuring drug levels, but semi-
quantitative data indicate the drug was localized to
some extent (though perhaps not as much as expected)
in adipose tissue. However, activity as an antilipolytic agent was not enhanced. Apparently the difluoro

substitution did not adequately retard the oxidative
deamination of this compound. Furthermore the subcellular localization of the drug may not have been proper
for optimum antilipolytic action. Nicotinic acid may
act on the adenyl cyclase contained in the cell
membrane (22), whereas the difluoro substituted amine
may be predominantly localized inside the adipocyte
such that the concentration at the precise subcellular
site of action is not higher than that of nicotinic
acid. Whatever the explanation, the application of pK_a
reduction through difluoro substitution was not successful in this instance.

We have also studied β,β-difluoro substituted
N-cyclopropyl-phenethylamines, compounds which are
irreversible inhibitors of monoamine oxidase. Monoamine oxidase inhibitors have been used clinically to
treat mental depression and hypertension. Another
pharmacologic action of monoamine oxidase inhibitors
that has potential clinical application involves
adipose tissue as a target organ.

Stock and Westermann (23) have shown that monoamine oxidase inhibitors elevate norepinephrine levels
in adipose tissue. The norepinephrine is presumably
contained in adrenergic nerve endings that control the
cyclic AMP-activated lipase in the fat cells and consequently the rate of lipolysis. Exposure to cold leads
to increased mobilization of fat depots and elevation
of plasma free fatty acid levels. In rats whose adipose tissue norepinephrine levels had been increased by
monoamine oxidase inhibition, exposure to cold produced
a greater increase in plasma free fatty acids than
occurred in control rats (23).

Of course the use of monoamine oxidase inhibitors
would have effects in the brain and other organs inner-
vated by the adrenergic system as well, so one could
not selectively affect fat mobilization in this way.
We wondered if the reduction of pK_a values by β,β-
difluoro substitution might enhance the localization of
the monoamine oxidase inhibitor in fat and thus make
its effects somewhat selective, maximizing enhanced
lipid mobilization while minimizing CNS effects and
effects on the cardiovascular system.

The possible use of lipolytic agents in treating
obesity by mobilizing fat depots has long been
considered. The idea of using a drug to enhance
endogenous lipolytic stimuli as opposed to a drug that
directly caused lipolysis seemed especially interesting.
Thus we have prepared β,β-difluoro-N-cyclopropyl-p-
chlorophenethylamine and studied its properties as an
inhibitor of monoamine oxidase. N-Cyclopropyl amines

of this sort have long been known to inhibit monoamine
oxidase irreversibly (24). The β,β-difluoro compounds,
like the parent N-cyclopropylamines, were found to
inhibit the enzyme in a manner that was not reversible
by dialysis.

The β,β-difluoro derivative had the expected lower
pK_a value (Table X). However when we injected the
drugs into rats and measured monoamine oxidase inhibition in various tissues, we did not observe any selec-
tivity in the inhibition of the enzyme in fat (Table
XI). Again methodologic difficulties have kept us from
obtaining adequate data on distribution of the drugs,
but we can assume from data on similar drugs that the
difluoro compound would localize in fat. Why then was
it not a more effective inhibitor of monoamine oxidase
in this tissue? Possibly the monoamine oxidase in the
fat tissue is located within adrenergic neurons to a
much greater extent than in adipocytes, whereas the
drug is localized chiefly in the adipocytes. For whatever reason, selective inhibition of adipose tissue
monoamine oxidase was not achieved.

The β,β-difluoro substituted monoamine oxidase
inhibitors may nonetheless have interesting uses as
pharmacological tools. The fluoro substituents might
permit NMR studies that would reveal something of the
nature of the active site of monoamine oxidase to which
the inhibitor is irreversibly bound.

TABLE X

*Ionization of N-Cyclopropyl-4-chlorophenethylamine
and its β,β-Difluoro Derivative*

Compound	pK_a	At pH 7.4, % as	
		RNH_2	RNH_3+
Cl—⟨phenyl⟩—CH_2CH_2NH—◁	8.3	20	80
Cl—⟨phenyl⟩—CF_2CH_2NH—◁	5.5	99	1

TABLE XI

Inhibition of Monoamine Oxidase in Rat Tissues *In Vivo*

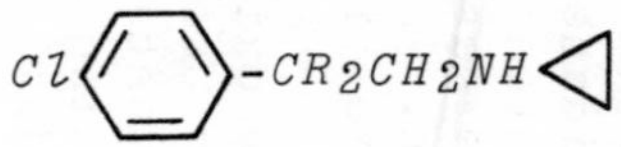

$$Cl-\langle\text{C}_6\text{H}_4\rangle-CR_2CH_2NH-\triangleright$$

Tissue	Group	Experiment 1. R = H		Experiment 2. R = F	
		MAO Activity	% Inhibition	MAO Activity	% Inhibition
Liver	Control	513 ± 21		450 ± 23	
	Drug	213 ± 15	59	137 ± 6	69
Brain	Control	107 ± 2		114 ± 1	
	Drug	52 ± 1	51	42 ± 1	63
Adrenals	Control	106 ± 5		109 ± 5	
	Drug	40 ± 1	62	42 ± 2	62
Kidneys	Control	57 ± 1		51 ± 1	
	Drug	22 ± 1	61	20 ± 1	60
Heart	Control	25 ± 1		25 ± 2	
	Drug	10 ± 1	60	13 ± 1	46
Fat	Control	17 ± 1		24 ± 1	
	Drug	7 ± 1	59	11 ± 1	55

Drugs were injected i.p. at 50 mg/kg 1 hour before the rats were killed. Monoamine oxidase activity in tissue homogenates was assayed with ^{14}C-tryptamine as substrate.

Summary

Substitution of one or two fluorine atoms on the
β carbon of arylalkylamine drugs markedly reduced the
pK_a of the amines. Monofluoro substitution reduced the
pK_a more than 1 pH unit, but the pK_a was still above
physiological pH. In general, monofluoro substitution
led to small changes or no change in pharmacologic
properties of the drugs. Difluoro substitution further
reduced the pK_a to below physiological pH. Thus at
physiological pH the difluoro compounds are mostly
neutral whereas parent amines are nearly completely
cationic. Presumably as a result of this effect on
pK_a, difluoro substitution markedly altered the
properties of amphetamine and phenethylamines.

As a result of difluoro substitution, amphetamine
became a better substrate for microsomal deaminases
whereas phenethylamine became a poorer substrate for
mitochondrial monoamine oxidase in vitro; amphetamine
was bound less readily to albumin; amphetamine and
phenethylamine became better substrates for lung
N-methyltransferase; amphetamine, phenethylamine and
p-chloroamphetamine were distributed differently in
mouse and rat tissues, localizing least in fat without
difluoro substitution but most in fat with that
substitution; amphetamine metabolism in the rat was
shifted from para-hydroxylation to oxidative deamination; the metabolism of difluoroamphetamine but not
amphetamine in rats was accelerated by phenobarbital
pretreatment to induce microsomal enzymes in liver;
amphetamine hyperthermia in rats and locomotor stimulation in mice decreased proportional to drug levels
in brain; amphetamine lost the ability to deplete
heart and brain norepinephrine even at high doses;
p-chloroamphetamine's ability to lower brain serotonin
levels initially was not altered except to the extent
that drug levels in brain were lower, but the neurotoxic effect of p-chloroamphetamine on serotonin
neurons was lost.

These results illustrate the importance of ionization in the interaction of amine drugs with biological
macromolecules both in vitro and in vivo. Fluorine
substituted near the nitrogen, because of its strong
electronegativity, can reduce the basicity of the amino
group to the extent that its ionization at physiological pH is sharply reduced. Fluorine substitution thus
alters many of the pharmacologic properties of such
amine drugs.

Acknowledgments

We thank Drs. W. S. Marshall, W. N. Shaw, and R. E. Toomey for their collaboration in the studies on the pyridine compounds. We also are grateful for the technical assistance of Kenneth L. Hauser in the synthesis of fluorinated derivatives and of Harold D. Snoddy, Betty W. Roush, and John C. Baker in the biological studies.

Literature Cited

1. Pauling, L. "The Nature of the Chemical Bond" p. 93 & 260, 3rd ed., Cornell University Press, Ithaca, 1960.
2. Loncrini, D. F. and Filler, R., Advances in Fluorine Chemistry (1970) 6, 43-67.
3. Henne, A. L. and Stewart, J. J., J. Am. Chem. Soc. (1955) 77, 1901-1902.
4. Lewis, G. P., Brit. J. Pharmacol. (1954) 9, 488-493.
5. Vree, T. B., Muskens, A. Th. J. M., and van Rossum, J. M., J. Pharm. Pharmacol. (1969) 21, 774-775.
6. Fuller, R. W. and Roush, B. W., Res. Comm. Chem. Pathol. Pharmacol. (1975) 10, 735-738.
7. Fuller, R. W., Molloy, B. B., and Parli, C. J. In "Psychopharmacology, Sexual Disorders, and Drug Abuse" pp. 615-624, Avicenum Press, Prague, 1973.
8. Fuller, R. W., Parli, C. J., and Molloy, B. B. Biochem. Pharmacol. (1973) 22, 2059-2061.
9. Fuller, R. W., Molloy, B. B., Roush, B. W. and Hauser, K. L., Biochem. Pharmacol. (1972) 21, 1299-1307.
10. Dring, L. G., Smith, R. L., and Williams, R. T., Biochem. J. (1970) 116, 425-435.
11. Gessa, G. L., Clay, G. A., and Brodie, B. B., Life Sci. (1969) 8, 135-141.
12. Fuller, R. W., Shaw, W. N., and Molloy, B. B., Arch. Int. Pharmacodyn. (1972) 199, 266-271.
13. Moore, K. E., J. Pharmacol. Exptl. Therap., (1963) 142, 6-12.
14. Fuller, R. W., Snoddy, H. D., Molloy, B. B., and Hauser, K. L., Psychopharmacologia (1973) 28, 205-212.
15. Sanders-Bush, E., Bushing, J. A., and Sulser, F., Eur. J. Pharmacol., (1972) 20, 385-388.
16. Fuller, R. W. and Snoddy, H. D., Neuropharmacol., (1974) 13, 85-90.

17.   Harvey, J. A., McMaster, S. E., and Yunger, L. M.,
      Science (1975) <u>187</u>, 841-843.
18.   Fuller, R. W., Snoddy, H. D., and Molloy, B. B.
      J. Pharmacol. Exptl. Therap. (1973) <u>184</u>, 278-284.
19.   Fuller, R. W. and Molloy, B. B., Adv. Biochem.
      Psychopharmacol. (1974) <u>10</u>, 195-205.
20.   Fuller, R. W., Perry, K. W., and Molloy, B. B.,
      Eur. J. Pharmacol. (1975) <u>33</u>, 119-124.
21.   Levy, R. I. and Frederickson, D. S., Postgrad.
      Med. (1970) <u>47</u>, 130-136.
22.   Peterson, M. J., Hillman, C. C. and Ashmore, J.,
      Mol. Pharmacol. (1968) <u>4</u>, 1-9.
23.   Stock, K. and Westermann, E. O., J. Lipid Res.,
      (1963) <u>4</u>, 297-304.
24.   Mills, J., Kattau, R., Slater, I. H. and Fuller,
      R. W., J. Med. Chem. (1968) <u>11</u>, 95-97.

Question and Answer Period

Q.   Have you prepared any ring-hydroxylated compounds?

A.   *No. We particularly wanted β,β-difluoro-p-
     hydroxyamphetamine but have not been able to
     synthesize it. There may be a problem of stabil-
     ity with compounds having both a hydroxyl and a
     -CF₂R group on the ring.*

Q.   Is fluoride ion a metabolite of these difluoro
     compounds?

A.   *We do not know whether any of the fluorine is
     removed metabolically.*

Q.   About the possibility of changing the basicity of
     the nitrogen in the difluoro compounds--have you
     looked at the effect of the β,β-difluoro substi-
     tuent on transport across biological membranes?

A.   *Mostly in indirect ways. I think the large change
     in tissue distribution produced by the β,β-
     difluoro substitution is a manifestation of
     altered transport across various biological mem-
     branes in the complex system of the whole animal.
     Dr. David Wong has looked at the ability of β,β-
     difluoro-p-chloroamphetamine to inhibit the trans-
     port of monoamines across synaptosomal membranes
     in vitro; the difluoro compound is less effective
     as an inhibitor than is p-chloroamphetamine
     itself.*

Q. Do you know what the products are from oxidative
    deamination of the β,β-difluoro compounds?

A. *Dr. John Parli has isolated the oxime, the ketone,
    and the alcohol as* in vitro *metabolites of β,β-
    difluoroamphetamine. He has found that the oxime
    of difluorophenylacetone is formed as a metabolite*
    in vivo *in rats and is excreted in urine in free
    and conjugated form.*

Metabolic and Transport Studies with Deoxyfluoro-monosaccharides

N. F. TAYLOR, A. ROMASCHIN, and D. SMITH

Department of Chemistry, University of Windsor, Ontario N9B 3P4 Canada

The rationale for the synthesis and use of fluorocarbohydrates and related compounds as probes for the study of enzyme specifity, carbohydrate metabolism and transport in biological systems has been elaborated in a previous symposium (1). Such compounds have now been used to study carbohydrate metabolism in yeast cells (2), Ps. fluorescens (3) and E. coli (4); enzyme specificity of glycerol kinase (5), yeast hexotinase (6), phosphoglucomutase and UDPG pyrophosphorylase (7) and glycerol-3-phosphate dehydrogenase (8); carbohydrate transport in hamster intestine (9) and the human erythrocyte (10), (11). The wide range of synthetic fluorinated carbohydrates and related compounds now available has been extensively reviewed (12). As will be evident from our recent studies, however, many detailed biochemical studies will be limited until the introduction of ^{14}C and/or ^{3}H into fluorocarbohydrates has been accomplished.

The Transport of D-Glucose Across the Human Erythrocyte Membrane.

The question of the exact nature of the carrier protein(s) and translocation mechanism for the transport of D-glucose across the erythrocyte membrane is still debated (13). However, the saturation kinetics obtained for D-glucose and glucose analogues are in accordance with a facilitated transport mechanism which allows binding of the sugar molecule to one or more sites of a receptor protein in the membrane. A comprehensive study of the specificity of this binding has been undertaken by a number of workers (14), (15), (16), (17), In particular the comparative inhibition and comparative transport studies of D-glucose with a number of monodeoxy-D-glucoses and monodeoxymonofluoro-

D-glucoses provide kinetic parameters which permit
assignment of hydrogen bonds between specific oxygen
atoms in D-glucose and receptor sites in the carrier
protein. Thus using the optical method of Sen and
Widdas (15) and the simplified rate equation (18) for
the exit of glucose from pre-loaded erythrocytes, we
have shown (11) that replacing the oxygen function at
C_3 of D-glucose by fluorine to give 3-deoxy-3-fluoro-
D-glucose does not significantly change the half-sat-
uration constant (K_x) for the carrier protein (Table 1).
In contrast 3-deoxy-D-glucose has lost this ability to
hydrogen bond at C_3 and consequently has a lower affinity for the carrier protein (higher K_x value). In
addition the K_x value for 5-thio-D-glucose (Taylor, N.F.
& Gagneja, G.L. unpublished result) suggests that the
ring oxygen at C_5 of D-glucose is also involved with
hydrogen bonding (Table 1).

Table 1. K_x and V_{max} values of D-glucose
            and derivatives at 37°

Sugar	K_x (mM)	V_{max} (mmol. Litre^{-1} min^{-1})
D-Glucose	3.9	640
3-Deoxy-3-fluoro-D-glucose	2.3	600
3-Deoxy-D-glucose	15.3	340
5-Thio-D-glucose	15.0	500

Kinetic parameters were determined as previously
described (11).

These results are in agreement with Barnett et al. (10)
who showed by inhibition studies that unlike 3-deoxy-
D-glucose, 3-deoxy-3-fluoro-D-glucose binds to the
carrier protein as well as D-glucose. Furthermore,
the importance of the β-orientation of -OH at C_1 in
D-glucose was suggested by the high Ki values for
1-deoxy-D-glucose and α-D-glucosylfluoride and the low
Ki value for β-D-glucosylfluoride. These transport
and inhibition studies provide evidence for the
proposal by Kahlenburg and Dolansky (19) that the
oxygen functions located at C_1, C_3 and the ring oxygen
at C_5 of the C1-conformation of β-D-glucopyranose
(Figure 1), are considered to be necessary for effective hydrogen bonding of D-glucose to the transport
protein. This model agrees with the detection of three
different receptor groups (- NH_2, - SH and imidazole)
on the protein associated with glucose transport (20)
and is also consistent with a recently proposed model

(21) for the mode of action of cytochalasin B (22)
inhibition of glucose transport in the human erythrocyte. Thus a Drieding molecular model (Figure 2) of
cytochalasin B (I) reveals an almost identical spatial
distribution of the four oxygen atoms located at Cl,
C19, C18 and C4 to those located at C5, Cl, C2 and C3
of the Cl-conformation of β-D-glycopyranose (Figure 1).
At least three of these sites at R, Rl and R3 (Figure
2) are implicated in hydrogen bonding to the carrier
protein for D-glucose and partially explain why
cytochalasin B is a competitive inhibitor of D-glucose
transport in the human erythrocyte with Ki, 1.2 x 10^{-7}
M (21).

A further point of interest resides in the fact
that when whole red blood cells are incubated with 3-
deoxy-3-fluoro-D-glucose for 24 hrs at 37°C a small
but significant release of free fluoride anion occurs
(Halton, D. & Taylor, N.F. unpublished results) using
200mM 3-deoxy-3-fluoroglucose C-F cleavage reaches a
maximum (∿ 1%) after 24 hours incubation (Figure 3).
Unlike the controls such cells completely lose their
ability to transport glucose. The sigmoidal curve
indicates multiple kinetics for the mechanism of
fluoride release and one possible explanation for these
results would be the concommittant co-valent attachment
of the glucose residue to one of the receptor sites of
the carrier protein associated with glucose transport.
Such a mechanism is shown (Figure 4) in which 3-deoxy-
3-fluoroglucose (II) hydrogen bonds to the receptor
protein, eliminates HF to produce the epoxide (III)
which is attacked by the nucleophilic protein to produce (IV). We have recently synthesised 3H-C-(3)-3-
deoxy-3-fluoro-D-glucose (Lopes, D. & Taylor, N.F.
unpublished results) in order to establish whether
glucosylation of a membrane protein has in fact occurred. Such a reaction may permit isolation of the
carrier protein for D-glucose.

Microbial Metabolism of Deoxyfluoro-D-glucoses

Our previous studies (3) demonstrated that 3-deoxy-
fluoro-D-glucose (II) is metabolised by whole resting
cells of Ps. fluorescens, with retention of the C-F
bond, to produce 3-deoxy-3-fluoro-D-gluconic acid (V).
Cell-free extracts of this organism oxidised 3FG
further to 3-deoxy-3-fluoro-2-keto-D-gluconic acid (VI).
It has also been shown that the same enzymes that
oxidise D-glucose (glucose oxidase and gluconate dehydrogenase) oxidise (II) and that (II) and (V) are
competitive inhibitors of gluconokinase (23). In
order to study further the specificity of these enzymes

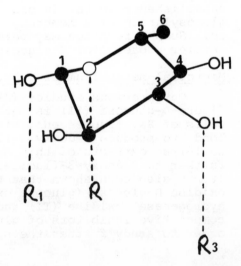

(I)

Figure 1. *Stereospecific binding sites of β-D-glucopyranose in the C1-conformation to a transport protein . . . Hydrogen bonds. R, R1 and R3 represent receptor sites on the transport protein.*

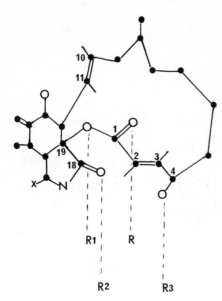

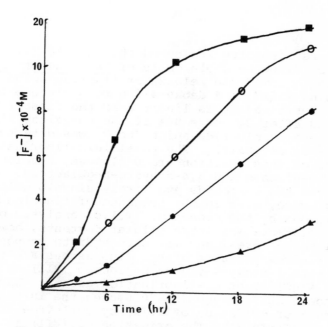

Figure 2. *Model of cytochalasin B. C_4 and C_{19} are in the R-conformation.* ● *Carbon atoms.* ○ *oxygen functions. . . . Hydrogen bonds.* $X = -CH_2Ph.$ *R, R1, R2 and R3 represent receptor sites on the transport protein.*

Figure 3. *Fluoride ion released after incubation of human erythrocytes at 37° with different concentrations of 3-deoxy-3-fluoro-D-glucose:* ▲ *50 mM,* ● *100 mM,* ○ *150 mM,* ■ *200 mM. Fluoride ion determinations were by the fluoride electrode (26) and the preparation of the erythrocytes as previously described (11). The initial control fluoride ion concentration was $1.0 \times 10^{-4}M$.*

towards other deoxyfluoro-D-glucoses we have examined
the biochemical effects of the isomeric 4-deoxy-4-
fluoro-D-glucose (VII) (24) on whole resting cells and
cell-free extracts of Ps. fluorescens (Hill, L. &
Taylor, N.F. unpublished results).

(II)   (V)   (VI)   (VII)

Warburg respirometry indicated that unlike (II) (VII)
is not oxidised by whole cells of Ps. fluorescens but
there is an immediate release of fluoride anion (Figure
5). Using 2.5mM of 4-deoxy-4-fluoro-D-glucose the
rate of C-F cleavage is linear over the first four
hours and after 24 hours 94% of the co-valent fluorine
is released as fluoride anion (no F⁻ was detected in
the absence of cells). The possible defluorinated
products of this reaction are D-glucose, D-galactose,
4-deoxy-D-glucose and 3,4-anhydro-D-galactose. Only
the latter two compounds are likely since D-galactose
and D-glucose, even in the presence of fluoride anion,
are oxidised by the organism. T.l.c. analysis of the
cell superatants and intracellular contents, however,
has failed to reveal any new carbohydrate components.
When cell-free extracts of Ps. fluorescens are challenged with 4-deoxy-4-fluoro-D-glucose (VII) we find
that no significant de-fluorination occurs. Thus in
the concentration range 2 - 20 μmoles the initial rate
and extent of oxidation of (VII) by cell-free extracts
is shown (Table 2). The oxidation of (VII) is complete within 2 hours and 2 g atoms of oxygen/mole
of substrate are consumed.

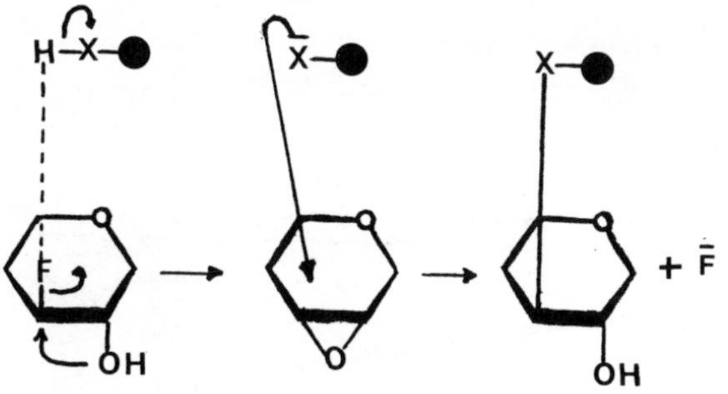

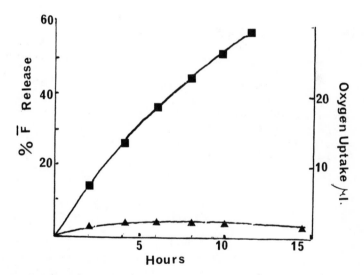

Figure 4. Possible mechanism of fluoride release from 3-deoxy-3-fluoro-D-glucose. ● *protein X = N or S.*

(II)   (III)   (IV)

Figure 5. Release of fluoride anion from 2.5 mM 4-deoxy-4-fluoro-D-glucose by whole cells of Ps. fluorescens. ■ Fluoride anion. ▲ Oxygen uptake. Six Warburg flasks were used. Each flask contained 5 μmoles 4-deoxy-4-fluoro-D-glucose, 20 mg dry wt cells and 0.67M phosphate buffer to 2.0 ml in main well. Flask contents were incubated at 30°C with shaking. At time intervals the contents of each flask were centrifuged at 4040 g for 20 min and fluoride ion determinations made with a fluoride electrode (26) on the supernatants. Warburg conditions for oxygen uptake: 30°C, reaction vol 2.0 ml. Flask contained 1.0 ml 0.67M phosphate buffer and 0.5 ml 5 μmoles of 4-deoxy-4-fluoro-D-glucose in main well and 0.2 ml 20% KOH aq. in center well. Reaction was initiated by tipping 0.5-ml cell suspension (24 mg dry wt) in 0.67M phosphate buffer from side arm. Endogenous respiration subtracted (1434 μl in 480 min).

Table 2.   Oxidation of 4-deoxy-4-fluoro-D-glucose by
           cell-free extracts of Ps. fluorescens.

Amount of 4-fluoroglucose μmoles	Rate of oxidation μmoles O_2/ hr/mg protein	Net μl oxygen consumed	Moles O_2/ mole of substrate oxidised
2.0	0.07	50	1.12
5.0	0.18	110	0.99
10.0	0.26	225	1.00
20.0	0.34	440	0.96
5.0 (glucose)	0.38	115	1.03

Warburg conditions: 30°C, gas phase, air.   Reaction
volume, 2.0 ml.   Each flask contained 37 mg cell-free
extract protein, 1 μmole NAD, 0.67 M phosphate buffer,
pH, 7.0 made up to 1.5 ml in the main compartment.
Side arm contained 0.5 ml of substrate and the centre
well 0.2 ml 20% KOHaq and paper wick.   The reaction
was initiated by typing contents from the side arm.
Endogenous respiration (317 μl oxygen in 150 min)
subtracted.

        After oxidation of (VII) was complete, silica gel
t.l.c. analysis (EtOAC:ACOH:H_2O 3:3:1) of the cell-free
extract revealed the absence of (VII)(R_F, 0.6) and the
presence of a new component (R_F 0.45).   By analogy
with the established metabolic pathway of glucose (27)
and 3-deoxy-3-fluoro-D-glucose (23) in this organism,
these results are consistent with the two step oxidation of (VII) by glucose oxidase and gluconate dehydrogenase to 4-deoxy-4-fluoro-D-gluconic acid (VIII) and
4-deoxy-4-fluoro-2-keto-D-gluconic acid (IX) respectively.

(VII)   (VIII)   (IX)

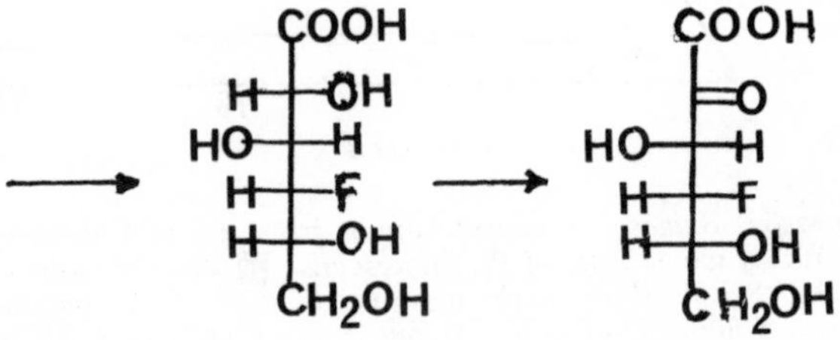

Isolation and characterisation of (IX) is currently
being investigated. The extensive defluorination of
(VII) by whole cells of Ps. fluorescens and the retention of the C-F bond on treatment of (VII) with cell-
free extracts suggest that C-F cleavage occurs at the
cell-wall/membrane level of the organism. The mode of
uptake of D-glucose by Ps. fluorescens is not known.
However, in a closely related species Ps. aeruginosa,
it has been shown (28) that although the phosphoenol-
pyruvatephosphotransferase system (29) is not involved,
the transport of D-glucose is energy dependant and,
therefore, likely to have a carrier protein system.
A similar glucose transport system may be present in
Ps. fluorescens and the failure of the whole cell to
oxidise 4-deoxy-4-fluoro-D-glucose (VII) may be due to
a stereospecific reaction of (VII)with a carrier
protein system in which co-valently bonded fluorine
is released as fluoride ion. Such a reaction would
permit the sugar residue (at C_4) to become attached to
protein in the membrane and account for the fact that
we are unable to detect any metabolite either outside
or inside the cell during the incubation period. Some
support for this possibility is provided by (a) the
fact that when whole cells are pre-incubated with
2.5mM 4-deoxy-4-fluoro-D-glucose for 12 hours and
challenged with 2 - 8mM glucose significant inhibition
of the rate of respiration occurs (Figure 6). This
would be consistent with a blocked glucose transport
site(s) although as the glucose concentration is
increased to 20 μmoles some recovery of respiration is
apparent. The possibility that a small undetectable
amount of 4-deoxy-4-fluoro-D-glucose or a non-oxidis-
able fluorinated metabolite is inhibiting glucose
metabolism and/or transport is of course not excluded
by these results. (b) Ps. fluorescens is unable to
grow on a mineral salts medium with 4-deoxy-4-fluoro-
glucose (VII) as a sole carbon source although fluorine
ion is released into the medium. Several examples are
known where the release of fluoride from a C-F compound by a bacterium allows the free non-fluorinated
fragment to serve as a source of carbon and energy for
growth. Thus a Pseudomonad has been isolated which
grows on fluoroacetate as a result of C-F cleavage (30).
Similarly, a Pseudomonad has been isolated which will
grow on monofluorocitrate after fluoride release (31).
The inability of 4-deoxy-4-fluoroglucose (VII), to act
as a carbon source for Ps. fluorescens, despite C-F
cleavage, may be due therefore, to the attachment of
the sugar residue to some cellular component.
      It is expected that the synthesis of ^{14}C-(1)-4-

Figure 6. *Oxidation of D-glucose by resting whole cells of* Ps. fluorescens *after pre-incubation with 2.5 mM 4-deoxy-4-fluoro-D-glucose. Pre-incubation conditions: 30°C, reaction volume 2.0 ml, time 12 hr. Each Warburg flask contained 1.0 ml 0.67M phosphate buffer, 0.5 ml 5 μmoles of 4-deoxy-4-fluoro-D-glucose or 5 μmoles D-glucose in main well and 0.5-ml cell suspension (28 mg dry wt) in 0.67M phosphate buffer in side arm. Pre-incubation was initiated by tipping contents from side arm. Oxidation of added D-glucose, Warburg conditions: 30°C, reaction volume 2.5 ml, gas phase, air. After preincubation period, 0.5 ml of μmole quantities of D-glucose were added from the side arm, 0.3 ml 20% KOHaq added to the center well. Endogenous respiration subtracted (879 μl in 240 min).*
■ *Average oxidation of 5, 10, 15 and 20 μmoles glucose after pre-incubation with 5 μmole glucose.*
● *10 μmole,* △ *15 μmole,* ▲ *20 μmole glucose added after pre-incubation of cells with 2.5 mM 4-deoxy-4-fluoro-D-glucose.*

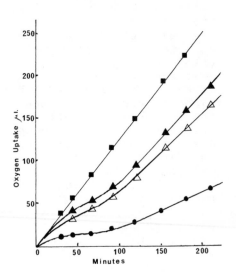

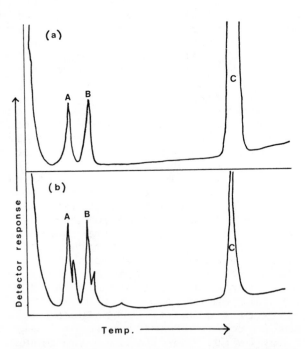

Figure 7. *(a) Temperature programmed gas chromatogram of locust haemolymph and (b) Locust fat body utilizing OV-17 (7.5%) phase at 170°C followed by a 4°C/min heating rate. Peaks: A = α-D-glucose, B = β-D-glucose, C = trehalose.*

fluoro-D-glucose, based on a Kiliani extension of
3-deoxy-3-fluoro-D-arabinose (32) with Na^{14}CN, will
allow us to ascertain the actual metabolic fate of
4-deoxy-4-fluoro-D-glucose (VII) and its defluorinated
product. In order to examine the bacterial specificity
of this C-F cleavage we have recently examined the
biochemical effects of (VII) on E. coli (ATCC 11775)
and shown that no significant C-F cleavage by whole
cells or cell-free extracts occurs. A small but
significant uptake of (VII) is observed (0.06 mg/mg
dry weight of bacteria). Using another strain of E.
coli (ATCC 14948) we have also demonstrated (Louie,
Li-Yu & Taylor, N.F. unpublished results) that (VII)
prevents utilization of lactose in this organism by
uncompetitive inhibition of the induction of β-
galactosidase. Our results are similar to those reported for the effects of 3-deoxy-3-fluoro-D-glucose
on E. coli (4).

Toxicity of 3-deoxy-4-fluoro-D-glucose in Locusta migratoria and Schistocerca gregaria

Although 3-deoxy-3-fluoro-D-glucose (II) displays
several physiological and biochemical effects in rats
(33), it is not toxic. Coupled with the fact that
rats excrete large quantities of unchanged (II) via
urine and faeces and also the relatively large quantities of (II) (5g/Kg body weight) necessary to evoke
a biochemical response it was considered to be of some
interest to study an animal organism which required a
smaller dosage of (II) and retained water to a greater
extent. Two closely related African locust species,
Schistocerca gregaria and Locusta migratoria were
chosen for this purpose. In both 10-14 day adult
insects (II) was toxic (LD$_{50}$ 5mg/g locust tissue) when
injected into the haemocoel. This compound was also
found to be toxic when orally ingested. Toxicity was
evidenced by progressive loss of motoractivity with
death occurring between 30 hours to 4 days. These
symptoms suggested that a slow metabolic poisoning was
occurring.
    Using the gas chromatographic procedure outlined
by Ford and Candy (34) for obtaining a carbohydrate
scan, levels of various steady state neutral carbohydrate metabolites were assayed from various locust
tissues (Figure 7). It was found that levels of (II)
disappeared rapidly from hemolymph, fat body and flight
muscle within two hours of injection. Analysis of
excreta indicated that a significant portion of the
injected dose was lost. Gas chromatographic results

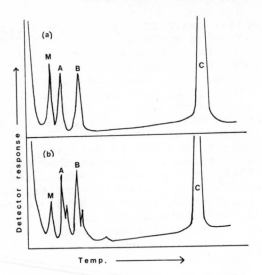

Figure 8. (a) Temperature programmed gas chromatogram of Locust haemolymph after 3-deoxy-3-fluoro-D-glucose injections. M = metabolite, A = α-D-glucose, B = β-D-glucose, C = trehalose. (b) Locust fat body after similar treatment.

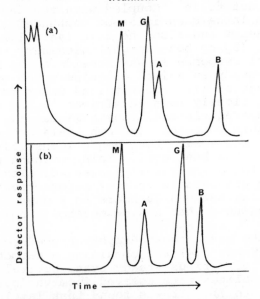

Figure 9. (a) Gas chromatogram (isothermal conditions) of locust hemolymph after 3-deoxy-d-fluoro-D-glucose and glucose administration using OV-17 stationary support. M = 3-deoxy-3-fluoro-glucitol, G = glucitol, A = α-D-glucose, B = β-D-glucose. (b) Using E301 stationary support.

indicated that a new metabolite appeared in the neutral
carbohydrate fraction following administration of (II).
This carbohydrate was shown to be the alditol of (II),
namely, 3-deoxy-3-fluoro-D-glucitol (X) by gas chromatographic and TLC comparison of the synthetically
made alditol (Lopes, D. and Taylor, N.F. unpublished
results).to the one obtained from locust tissue
extracts (Figures 8 and 9).   Furthermore, after insects
were poisoned with a 10.8 mg dose of (II) and 12 hours
later hemolymph glucose levels artifically raised by
a 10.8 mg dose of glucose the presence of 0.4 - 0.5
mg/100mg tissue of (X) and 0.3 - 0.6 mg/100mg tissue
D-glucitol was detected.   This is the first example of
sorbitol metabolism to be detected in either of these
species.   Under normal conditions sorbitol was not
detectable in the neutral carbohydrate fraction.   This
phenomenon may have been due to two possibilities,
(a) the rate of normal sorbitol utilization was much
greater than its rate of formation. (b) This was not
an important metabolic pathway in the locust and was
only detectable due to inhibition.
      Sorbitol metabolism has been reported in the silk
worm (35) and mosquito (36).   An extensive review of
polyols and their metabolism has been offered by
Touster and Shaw (37).   Our results suggested that the
pathway from glucose to fructose via sorbitol was
active in these locust species and that (II) blocked
the subsequent conversion of sorbitol to fructose
through the inhibitory action of (X) on the enzyme
sorbitol dehydrogenase (Figure 10).   A precursor
product   relationship is evident between (II) and (X)
in both hemolymph and fat body (Figure 11).   Initially
it was thought that the enzyme sorbitol dehydrogenase
was found in the hemolymph of the locust.   This is the
case in silkworm (35).   No activity, however, was
assayable in locust hemolymph using 0.1M sorbitol and
10mM nicotinamide adenine dinuceltide (NAD^+) as substrates.   Subsequent study revealed that the enzyme
activity was confined to the fat body.   Using about
10 fold purified enzyme preliminary results have
suggested that this enzyme is not very active in lo-
custs having a K_M for sorbitol in the order of 0.05M.
Preliminary results suggest that (X) is a competitive
inhibitor of the enzyme with a Ki∿0.1M.   These results
also suggest that the pathway of sorbitol metabolism
from glucose to fructose is not very important to the
total metabolism of the locust.   Therefore, it is
difficult to rationalize how the inhibition of such
a minor pathway can manifest toxic action.   It is
quite possible that some hitherto undetected

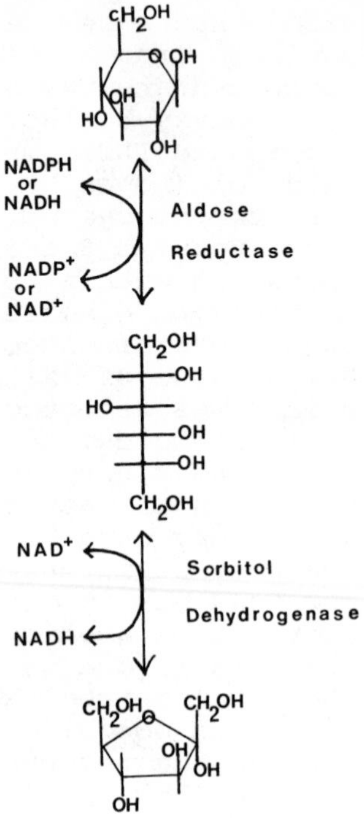

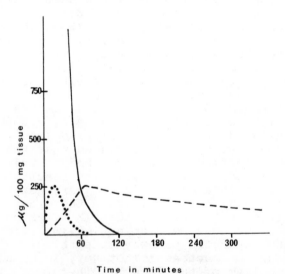

Figure 10.   Sorbitol conversion pathway.

Figure 11.   Levels of 3-deoxy-3-fluoro-D-glucose (II) and 3-deoxy-3-fluoro-D-glucitol (X) in locust hemolymph and fat body after a 3.6-mg injection of (II). ———— (II) in hemolymph ————(X) in hemolymph · · · · (II) in fat body.

metabolites are responsible for toxicity.

Chefurka (38), (39) has suggested a means of evaluating glucose metabolism in insects using $1-^{14}C$ and $6-^{14}C$ labelled D-glucose by examining the relative rates of $^{14}CO_2$ evolution. This technique may be a useful tool for monitoring any other changes in the pathways of glucose metabolism induced by (II) or its metabolites. Our studies on the mode of toxicity will be furthered when ^{14}C and/or ^{3}H-labelled fluoroglucoses become available but these preliminary results indicate that some fluorinated carbohydrates are not only toxic to insects but may also act as probes for the detection of unsuspected metabolic pathways.

Acknowledgment

This work is supported by the National Research Council of Canada.

Abstract

(a) The use of deoxyfluoro-D-glucose as probes for the study of glucose transport across the human erythrocyte membrane is discussed. These studies are also related to the mode of action of cytochalasin B inhibition of D-glucose across this membrane. (b) Evidence is presented to show that 4-deoxy-4-fluoro-D-glucose is not oxidised by whole resting cells of Ps. fluorescens (ATCC 12633) but an extensive release of fluoride anion occurs. With cell-free extracts of Ps. fluorescens however, 4-deoxy-4-fluoro-D-glucose is oxidised to the extent of 2 g atoms of oxygen/mole of substrate. Possible reasons for this C-F cleavage are discussed. (c) 3-Deoxy-3-fluoro-D-glucose is toxic to Locusta migratoria (L.D.$_{50}$ 5mg/g and is metabolised by an NAD-linked sorbitol dehydrogenase, which is located in the fat body of the insect, to 3-deoxy-3-fluoro-D-glucitol. The activity of the partially purified enzyme is low with Km, 0.05M for sorbitol. 3-Deoxy-3-fluoro-glucitol is a competitive inhibitor with Ki, 0.1M.

Literature Cited.

1. Ciba Fdn. Symposium: "Carbon-Fluorine Compounds: Chemistry, Biochemistry and Biological Activities." pp 1-417. Associated Scientific Publishers, New York, 1972.
2. Woodward, B., Taylor, N.F. & Brunt, R.V., Biochem. Pharmacol. (1971), 20 1071-1077.
3. Taylor, N.F., White, F.H. & Eisenthal, R.E., Biochem. Pharmacol. (1972), 21 347-353.
4. Miles, R.J. & S.J. J. Gen. Microbiol. (1973) 76 305-318.
5. Eisenthal, R.E., Harrison, R., Lloyd, W.J. & Taylor, N.F., Biochem. J. (1972), 130 199-205.
6. Bessel, E.M. & Thomas, P., Biochem. J. (1973) 131 843-850.
7. Wright, J.A., Taylor, N.F., Brunt, R.V. & Brownsey R.W., Chem. Commun. (1972) 691-692.
8. Silverman, J.B., Barbiatz, P.S., Mahajan, K.P., Buschek, J. & Foudy, T.P., Biochemistry (1975) 14 2252-2258.
9. Barnett, J.E.G., Ralph, A. & Monday, K.A., Biochem. J. (1970) 118 843-850.
10. Barnett, J.E.G., Holman, J.D. & Monday, K.A., Biochem. J. (1973) 131 211-221.
11. Riley, G.J. & Taylor, N.F., Biochem. J. (1973) 135 773-777.
12. Kent, P.W., Ciba Fdn. Symposium: "Carbon-Fluorine Compounds". pp 169-208.
13. Lieb, W.R. & Stein, W.D., Biochim. Biophys. Acta (1972), 265 187-207.
14. LeFevre, P.G., Pharmacol. Rev. (1961) 13 39-70.
15. Sen, A.K. & Widdas, W.F., J. Physiol (London) (1962) 160 392-403.
16. Lacko, L. & Burger, M. Biochem. J. (1962) 83 622-625.
17. Baker, G.F. & Widdas, W.F., J. Physiol. (London) (1972) 271 10p.
18. Miller, D.M. in "Red Cell Structure and Function". (Jamieson, G.A. & Greinwalt, T.A. Eds.) pp 240-292, J.B. Lippincott & Co. Philadelphia & Toronto, 1969.
19. Kahlenberg, A. & Dolansky, D., Canad. J. Biochem. (1972) 50 638-643.
20. Bloch, R.J. Biol. Chem. (1974) 249 1814-1822.
21. Taylor, N.F. & Gagneja, G.L., Proc. Canad. Fed. Biol. Soc. (1975) 18 p 4.
22. Aldridge, D.C., Armstrong, J.J. & Speake, R.N., J. Chem. Soc. (1967) 1667-1676.
23. Taylor, N.F., Hill, L. & Eisenthal, R.E., Canad. J. Biochem. (1975) 53 57-64.

24. Barford, A.D., Foster, A.B., Westwood, J.H., Hall, L., & Johnson, R.N., Carbohydr. Res. (1971), 19 49-61.
25. Umbreit, W.W., Burris, R.H. & Stauffer, J.F., "Manometric Techniques" pp 1-61. Burgess Publishing Co., Minneapolis, Minn. 1964.
26. Woodward, B., Taylor, N.F., & Brunt, R.V., Analyt. Biochem. (1971) 36 303-309.
27. Wood, W.A. & Schwerdt, R.F., J. Biol. Chem. (1953) 201 501-511.
28. Midgley, M. & Dawes, E.A., Biochem. J. (1973) 132 141 -
29. Kundig, W., Ghosh, S. & Roseman, S., Proc. Nat. Acad. Sci. (1964) 52 1067-1074.
30. Goldman, P. J. Biol. Chem. (1965) 240 3434-3438.
31. Kirk, K., Goldman, P., Biochem. J. (1970) 117 409-410.
32. Wright, J.A., and Taylor, N.F., Carbohydr. Res. (1967) 3 333-339.
33. Riley, G.J., Ph.D. Thesis University of Bath U.K. (1973).
34. Ford, W.C.L., Candy, D.J., Biochem. J. (1972) 130, 1101.
35. Faulkner, P., Biochem. J. (1958) 68 375.
36. Handel, E. Van, Comp. Biochem. Physiol. (1969) 29 1023.
37. Touster, O., Shaw, D.R.D., Physiol. Rev., (1962) 42 181.
38. Chefurka, W., in "The Physiology of Insecta"., Volume 2, pages 641-656, Academic Press Inc., New York, 1965.
39. Chefurka, W., Ela, R., and Robinson, J.R., J. Insect Physiol. (1970) 16 2137-2156.
40. Hall, L.D., Johnson, R.N., Foster, A.B. and Westwood, J.H., Canad. J. Chem. (1971) 49 236-240.
41. Lin, S. and Spudich, J.A., J. Biol. Chem. (1974) 249 5778-5783.

Q. Is the conformation of glucose, 3-deoxy-3-fluoro-
    glucose and 3-deoxy-glucose the same?

A. Although there is no direct experimental evidence,
    these three hexoses probably have the same pre-
    dominate Cl-conformation in aqueous solution with
    all the substituents equitorial. There is some
    evidence, however, that the introduction of fluorine
    can dictate which conformation the sugar ring will
    adopt. Thus Hall et al (40) have shown that in
    the case of the acetylated D-xylosyl fluoride
    derivatives the β-anomer adopts the unexpected
    conformation in which all the substituents are
    axially disposed.

Q. Does glucose bind with cytochalasin B?

A. No. The binding studies of Lin and Spudich (41)
    and our own kinetic results (21) in which cyto-
    chalasin B is shown to be a competitive inhibitor
    (Ki, 10^{-7}M) of glucose transport in the human
    erythrocyte suggest that cytochalasin B binds to a
    carrier protein and not to glucose.

Organic Fluorocompounds in Human Plasma: Prevalence and Characterization

W. S. GUY
Department of Basic Dental Sciences, University of Florida, Box J424, Gainsville, Fla. 32610

D. R. TAVES
Department of Pharmacology and Toxicology, University of Rochester, Rochester, N.Y. 14642

W. S. BREY, JR.
Department of Chemistry, University of Florida, Gainsville, Fla. 32611

Taves discovered that samples of his own blood serum contained two distinct forms of fluoride (1-4). Only one of these was exchangeable with radioactive fluoride. The other, non-exchangeable form was detectable as fluoride only when sample preparation included ashing. This paper is concerned with three aspects of this newly discovered, non-exchangeable form: 1) its prevalence in human plasma, 2) how its presence in human plasma affects the validity of certain earlier conclusions about the metabolic handling of the exchangeable form of fluoride, and 3) its chemical nature.

Preliminary work in this laboratory suggested that the non-exchangeable form was widespread in human plasma but did not exist in the plasma of other animals. Ashing increased the amount of fluoride an average of 1.6 ± 0.25 SD μM (range 0.4 - 3.0) in samples of plasma from 35 blood donors in Rochester, N.Y. (5). No such fluoride was detectable (above 0.3 μM) in blood serum from eleven different species of animal including horse, cow, guinea pig, chicken, rabbit, sheep, pig, turkey, mule and two types of monkey (6).

Standard methods for analysis of exchangeable fluoride in serum have in the past included ashing as a step in sample preparation (7). Taves showed that the amount of fluoride in serum that would mix with radioactive fluoride was only about one-tenth the amount generally thought to be present based on analyses using these older methods (4). When plasma samples from individuals living in cities having between 0.15 and 2.5 ppm fluoride in their water supply were analysed by these older methods, no differences were found between the averages for the different cities. This led to the conclusion that "homeostasis of body fluid fluoride content results with intake of fluoride up to and including that obtained through the use of water with a fluoride content of 2.5 ppm" (8). If the non-exchangeable form of fluoride predominated in these samples, differences in the exchangeable fluoride concentration would probably not have been apparent, and it would be unnecessary to postulate such rigorous homeostatic control mechanisms for fluoride.

In this study plasma samples were collected from a total of 106 individuals living in five different cities with between 0.1 and 5.6 ppm fluoride in their public water supply. These were analyzed for both forms of fluoride. In this way the relationship between exchangeable fluoride concentration in the plasma and the consumption of fluoride through drinking water was reevaluated, and the prevalence of the non-exchangeable form was further studied.

With respect to the chemical nature of the non-exchangeable form of fluoride several lines of evidence suggested that it was some sort of organic fluorocompound of intermediate polarity, tightly bound to plasma albumin in the blood. It migrated with albumin during electrophoresis of serum at pH nine (3) and was not ultrafilterable from serum (2). Attempts at direct extraction from plasma with solvents of low polarity like heptane, petroleum ether and ethyl ether were generally unsuccessful. Treatment of albumin solution (prepared by electrophoresis of plasma) with charcoal at pH three did remove the bound fluorine fraction. And finally, when plasma proteins were precipitated with methanol at low pH the fluorine fraction originally bound to albumin appeared in the methanol-water supernatant in a form which still required ashing to release fluorine as inorganic fluoride (5). Based on these considerations the non-exchangeable form of fluoride in human plasma is referred to as "organic fluorine" throughout the rest of this paper.

In order to further characterize the organic fluorine fraction, it was purified from 20 liters of pooled human plasma and characterized by fluorine nmr.

Materials and Methods

Analytical Methods. Values for organic fluorine were calculated by taking the difference between the amount of inorganic fluoride in ashed and unashed portions of the same material.

The following procedure was used to prepare ashed samples: 1) samples (sample size for plasma was 3 ml) were placed in platinum crucibles and mixed with 0.6 mmoles of low fluoride $MgCl_2$ and 0.1 mmoles of NaOH, 2) these were dried on a hotplate and then ashed (platinum lids in place) for 2-4 hr at 600° C in a muffle furnace which had been modified so that the chamber received a flow of air from outside the building (room air increased the blank and made it more variable), and 3) ashed samples were dissolved in 2 ml of 2.5 N H_2SO_4 and transferred to polystyrene diffusion dishes using 2 rinses with 1.5 ml of water.

The following procedure was used for separation of fluoride from both ashed and unashed samples: 1) samples (sample size for unashed plasma was 2 ml) were placed in diffusion dishes (Organ Culture Dishes, Falcon Plastics, Oxnard, Calif., absorbent removed, rinsed with water), acidified with 2 ml of 2.5 N H_2SO_4, and agitated with a gentle swirling action on a laboratory shaker for 30 min to remove CO_2; 2) for each sample the trapping solution (0.5 ml, 0.01 N NaOH + phenolthalein-p-nitrophenol indicator) was placed in a small polystyrene cup in the center-well of the diffusion dish, 1 drop of 10% Triton-X 100 was added to the sample to decrease surface tension and make the diffusion rate more uniform between samples containing plasma and those not, the lid with a small hole made near its lateral margin was sealed into place with petroleum jelly, 0.02 ml of 4% hexamethyldisiloxane (Dow Corning, Fluid 200, 0.65 cs, Midland, Mich.) in ethanol was injected through the hole in the lid into the sample, and the hole was sealed immediately with petroleum jelly and a strip of paraffin film; and 3) samples were diffused with gentle swirling for at least 6 hr, diffusion was terminated by breaking the seal, and trapping solutions were removed (the indicator color was checked at this point to insure that they were still alkaline) and dried in a vacuum oven (60° C, 26 in-Hg vacuum, in the presence of a NaOH desiccant).

Fluoride was determined by potentiometry with the fluoride electrode. The system used consisted of a fluoride electrode oriented in an inverted position (model 9409A, Orion Research Inc. Cambridge, Mass.), a calomel reference electrode (fiber type), a plastic vapor shield which just fitted over the bodies of both electrodes forming an enclosed sample chamber in which water-saturated tissue paper was placed above the sample to prevent evaporization of the sample, and a high impedence voltmeter (model 401, Orion).

Samples were prepared and read in the following way: 10 μl of 1 M HAc was drawn into a polyethylene micropipette (Beckman Micro Sampling Kit, Spinco Div., Beckman Inst. Co., Palo Alto, Calif.) and deposited into the cup containing the residue from the trapping solution after drying; the flexible tip of the micropipette was used to wash down the walls of the cup; and the solution was then transferred to the surface of the fluoride electrode and the reference electrode brought into position. Surfaces of the two electrodes were blotted dry between samples.

Samples were read in order of increasing expected concentration and sets of samples were read between bracketing calibration standards. These standards were used in two different ways during a run. First, they were flooded onto the electrode surfaces to equilibrate them to concentrations expected for samples and to make them uniform. This procedure permitted the analyst to take reasonably stable readings for samples within one minute. Secondly, they were used in 10 μl volumes for readings used in preparing the standard curve.

Values for identical samples (usually triplicates) were averaged and the average blank was subtracted from sample means. These were then divided by the average fractional recovery of fluoride (usually 90 to 95%) in standards treated the same way as the sample set.

Plasticware (Falcon Plastics) was used for all analytical procedures to avoid contamination by fluoride from glass. Liquid volume measurements were made with 1, 5 and 10 ml polystyrene pipettes and a polycarbonate volumetric flask (100 ml).

Reagents were purified to insure uniformly low blanks. Water was redistilled and deionized. Acetic acid and ammonia were redistilled. Fluoride contamination in $MgCl_2$ (analytical grade) was reduced by preparing a 1 M solution containing HCl to pH 1 and scrubbing with hexamethyldisiloxane vapor in a column through which the solution was continuously recycled. Following scrubbing the solution was boiled to one third volume to remove any residual volatile silicones and then made just basic with NH_4OH. Fluoride contamination in H_2SO_4 was reduced by repeated extractions of a 6.7 N solution with hexamethyldisiloxane and then boiling to one third volume to remove the residual silicone.

Buffered calibration standards were made from the same NaOH and HAc stock solutions as for samples.

The blanks for ashed samples ranged between 0.2 and 1.5 nmoles fluoride and were typically about 0.5 nmoles. The blanks were smaller for unashed samples; these ranged between 0.05 and 0.2 nmoles fluoride and were typically about 0.1 nmoles.

Factors affecting recovery of fluoride during diffusion were investigated with $^{18}F^-$ tracer. Recovery during diffusion was 97% after 80 min from 5 ml containing 2 ml of plasma. Increasing the acidity of the sample up to 5 N, the volume of the sample up to 7.5 ml, the amount of cold F^- up to 1 μmole, the amount of fluoride complexors up to 1 μmole of $Th(NO_3)_4$ had no material effect on the rate of fluoride diffusion. The absence of both plasma and detergent in the sample compartment markedly slowed the rate of diffusion. Not shaking the sample also slowed the rate of diffusion. Increasing the alkalinity of the trapping solution to 0.1 N increased the rate of diffusion but the lower concentration, 0.01 N, was required here to permit a lower ionic strength in the sample reading solution.

Overall recovery of added cold fluoride was measured. In samples containing neither plasma nor detergent the recovery after 6 hr diffusion averaged 93% and 95% for ashed and unashed samples, respectively. In samples containing plasma the recovery was 95% after 3 hr diffusion.

The degree to which fluorine from organic fluorocompounds could be fixed as inorganic fluoride by ashing varied from less than 1% for volatile compounds like p-aminobenzotrifluoride, m-hydroxybenzotrifluoride, benzyl fluoride and benzotrifluoride to over 80% for less volatile compounds like 5-fluorouracil, fluoroacetate and p-fluorophenylalanine.

Methods used here for separation of fluoride (diffusion at rm. temp.) (9) and its quantitation (fluoride electrode) (10) are considered to be quite specific for fluoride. One potentially important interference, however, was codiffusable organic acids which might partially neutralize the trapping solution and thus lower the pH of the buffered reading solution. Indeed, it was found that samples containing relatively large concentrations of acetic acid (e.g., fractions 2, 3 and 4 from step 4 in the purification system) completely neutralized the trap within a few hours. The significance of this problem in the analysis of fluoride in blood plasma was investigated in two ways. First, four samples of human plasma were allowed to diffuse for three weeks, and no change in the color of the phenolphthalein indicator in the trapping solution was observed. Secondly, samples containing the same plasma were diffused for different periods up to 158 hr and the apparent fluoride was determined. No changes were observed between samples which correlated with diffusion time.

The sensitivity of the analytical method was limited by the blank rather than the sensitivity of the instruments used. Reproducibility varied with the amount of fluoride being measured. The coefficient of variation averaged 55% in the low range (samples containing 0.25 to 0.75 nmoles F^-) and 6.6% in the high range (10-12 nmoles F^-).

Blood Plasma. Human plasma was obtained from blood banks in five cities. According to public records these cities had not changed the fluoride concentration of their public water supply for at least six years prior to obtaining the samples. Samples were received in individual polyethylene bags which were part of the Fenwall ACD blood collection system. In blood collection using this system 450 ml of blood is drawn into a bag containing 67.5 ml of anticoagulant acid citrate dextrose (ACD) solution. When the cells are removed the ACD solution remains in the plasma. Because of this dilution of plasma a correction factor of 1.3 was applied to values obtained here for the concentration of fluoride. The potential error in this factor was ± 0.1 because of variation between standard limits for hematocrit and minimum volume of the blood donation. Bovine blood was obtained at slaughter and mixed immediately with ACD solution in 1 liter polyethylene bottles.

Electrophoresis. A continuous flow electrophoretic separator (model FF-3, Brinkman Inst., Inc., Westbury, N.Y.) was employed. Sample flow rate was 2.3 ml/hr, buffer flow rate was 72 ml/hr, voltage was 0.67 kv, and current was 140 mamp. Separation took 19 hr. Plate separation was 1 mm and operating temperature was between 2 and 4° C. The buffer was 0.12% $(NH_4)_2CO_3$, made by bubbling CO_2 from dry ice into redistilled NH_4OH until the pH reached 9.0.

Purification System. Steps in the purification system are summarized in table I. In the first step one liter of plasma (pooled from 5-6 individuals) was dialysed in seamless cellulose tubing (1 in. diameter) against 20 liters of water at 4° C. The dialysate was changed twice at 24 hr intervals. In the second step dialysed plasma was freeze dried.

In the third step the dried powder from electrophoresis was extracted with methanol in a soxhelet extraction apparatus (model 6810 G, Ace Glass, Inc., Vineland, N.J.). Cellulose extraction thimbles (model 6812 G, Ace Glass) were soaked overnight in methanol. Operating conditions were 25° to 30° C under a vacuum of 24 in-Hg. Coolant for the condenser was 80% ethanol; inlet temperature was -10° to -20° C and outlet temperature was -10° to 0°. Two liters of methanol were refluxed through the apparatus for a period of 4 hr and approximately 400 ml were lost to evaporation during that period. Glass beads were placed in the flask to prevent bumping.

In the fourth step the residue from the methanol extract was fractionated according to the method described by Siakotos and Rouser (11) for separating lipid and non-lipid components. The method is based on liquid-liquid partition in a column containing a dextran gel (Sephadex G-25, coarse, beaded, Pharmacia Fine Chemicals, Inc., N.Y.). Four eluents are used: 1) 500 ml chloroform/methanol, 19/1, saturated with water, 2) 1000 ml of a mixture of 5 parts of chloroform/methanol, 19/1, and 1 part of glacial acetic acid, saturated with water, 3) 500 ml of a mixture of 5 parts chloroform/methanol, 19/1, and 1 part glacial acetic acid, saturated with water, and 4) 1000 ml of methanol/water, 1/1. Their method was modified for use here by increasing the column length to that attained by using a full 100 grams of dextran beads. Sample size corresponded to that from 2.5 liters of the original plasma.

In the fifth step the residues from eluents 2 and 3 from two runs of step four were combined, applied to a silicic acid column, and eluted by reverse flow with an exponential gradient of increasing amounts of methanol in chloroform. The column (model SR 25/45, 2.5 cm i.d. x 45 cm, Pharmacia) was filled to a height of 30 cm with silicic acid (Unisil, 100-200 mesh, Clarkson Chem. Co., Inc., Williamsport, Pa., heat activated at 110° C for 2 days) and was washed with a complete set of elution solvents before use. The gradient maker (model 5858, set 4, Ace Glass Co.) was filled with 1 liter of methanol in the upper chamber and 2 liters of chloroform in the lower. The flow rate was adjusted by the height of the solvent reservoirs to an average of 3 ml/min for the first liter of eluent. The sample had to be transferred to the column by repeated washings with chloroform because of its low solubility in this solvent. This usually required about 30 ml of chloroform total. Dead volume for the system as 90 ml. Fractions of 15 ml volume were collected in carefully cleaned glass tubes.

Table I

PROCEDURE FOR PURIFICATION
OF FLUOROCOMPOUNDS FROM BLOOD PLASMA

Fraction Treated	Treatment	Fraction Removed
blood plasma	step 1: exhaustive dialysis against distilled water	smaller, water-soluble components
plasma proteins & protein-bound substances in water solution	step 2: lyophilization	water
plasma proteins & protein-bound substances	step 3: methanol extraction--soxhelet, 25°C, 24 in-Hg vacuum	plasma proteins
plasma lipids	step 4: column chromatography-- liquid-liquid partition on Sephadex	lipids of low polarity and residual polar contaminants
polar lipids	step 5: column chromatography-- adsorption on silicic acid	unknown: several yellow fractions

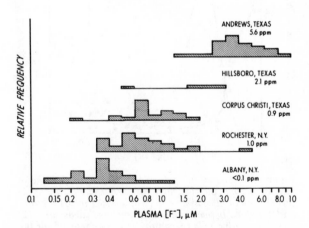

Figure 1. *Relationship between the concentration of fluoride in human plasma and the concentration of fluoride in the drinking water*

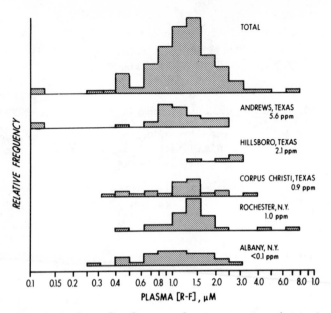

Figure 2. *Relationship between the concentration of organic fluorine in human plasma and the concentration of fluoride in the drinking water*

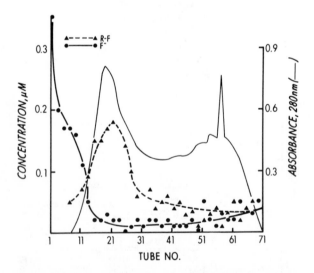

Figure 3. *Separation of fluoride and organic fluorine in human plasma by electrophoresis. A sample (about 45 ml) of human plasma was electrophoresed in pH 9 buffer and fractions between the sampling port (near tube 72) and the positive pole (near tube 1) were analyzed for the fluoride content of both ashed and unashed aliquots. Relative concentrations of proteins were estimated by absorbance at 280 nm.*

Tubing and fittings to the columns were polytetrafluoroethylene (supplied largely by Chromatronix, Inc., Berkeley, Calif.). All solvents were redistilled. Methanol and chloroform were ACS certified (Fisher Scientific Co.) and acetic acid was analytical reagent grade, U.S.P. (Mallinckrodt Chemical Works, St. Louis). Solvents were removed from samples in a flash evaporator.

NMR. The nmr spectrum was obtained on a Varian XL-100 spectrometer with Nicolet Technology Fourier Transform accessory. The sample was dissolved in an approximately 1/1 mixture of CH_3OH and $CDCl_3$ and spectra were run in a 5 mm tube. External referencing to $CFCl_3$ was used for the chemical shifts, and these are expressed with positive numbers to lower field (i.e., higher frequency). External lock was used. Typical conditions were a pulse length of 15 microseconds, a delay time between pulse cycles of 2.5 sec, and a time constant of -1 sec for exponential processing.

Results

Values for inorganic fluoride (F^-) and organic fluorine (R–F) in 106 plasma samples from humans living in five cities are shown in table II. These data show that the average fluoride concentration in plasma is directly related to the fluoride concentration in the water supply, and that the average organic fluorine concentration in plasma is not. No relationship between fluoride in plasma and organic fluorine in plasma was apparent by inspection of values for individual samples. The distributions of the values within cities are shown in figures 1 and 2. In both cases the distributions appear to be log normally distributed with only 3 or 4 individuals surprisingly deviant. In the cases of the two individuals with little or no apparent organic fluorine (figure 2, Andrews group, left margin), the inorganic fluoride levels were both in excess of 7 µM, making the difference measurement for organic fluorine difficult. The overall mean value for organic fluorine was 1.35 ± 0.85 SD µM.

Plasma was electrophoresed in an attempt to reproduce the findings of Taves (3) using plasma from another individual. Results shown in figure 3 closely match those found earlier in that a predominant form of organic fluorine appeared to migrate with albumin at pH 9, and in that organic and inorganic forms were clearly separated.

The recovery, mass balance and purification factors for steps in the purification system listed in table I are recorded in table III. These data show that about one-third of the original amount of organic fluorine in plasma is recovered in the major peak from silicic acid chromatography. Another third is accounted for in other fractions and the rest is not accounted for, presumably because of adsorption to surfaces of containers in

Table II

CONCENTRATION OF FLUORIDE (F⁻) AND ORGANIC FLUORINE
(R-F) IN BLOOD PLASMA SAMPLES FROM FIVE CITIES HAVING
DIFFERENT FLUORIDE CONCENTRATIONS IN THEIR WATER SUPPLY

City ([F⁻] in Water, ppm)	[F⁻] in Plasma[a], μM			[R-F] in Plasma[a,b], μM		
	Mean± SD(n)	Range	Diff.[c] P<.05	Mean± SD(n)	Range	Diff.[c,d] P<.05
Albany, N.Y. (<.1)	0.38± 0.21 (30)	0.14– 1.1		1.2± 0.6 (30)	0.3– 2.6	
			sig,			n.s.
Rochester, N.Y.(1.0)	0.89± 0.75 (30)	0.35– 4.2		1.6± 1.2 (30)	0.5– 6.8	
			n.s.			n.s.
Corpus Christi, Tex.(0.9)	1.0± 0.35 (12)	0.60– 1.7		1.3± 0.9 (12)	0.4– 3.9	
			sig.			n.s.
Hillsboro, Tex.(2.1)	1.9± 0.9 (4)	0.60– 2.6		2.3± 0.6 (4)	1.5– 2.8	
			sig.			sig.
Andrews, Tex. (5.6)	4.3± 1.8 (30)	1.4– 8.7		1.1± 0.5 (30)	0.1– 2.3	

[a]Each value used in the computation was the average of at least
    three replicate analyses and was corrected for dilution by
    ACD solution by multiplying it by 1.3.
[b]taken to be the difference between the amount of inorganic
    fluoride measured in ashed and unashed aliquots of the same
    sample
[c]by t-test assuming equal variance in each group
[d]The difference between Rochester and Andrews is statistically
    significant.

Table III

MASS BALANCE, RECOVERY AND PURIFICATION
FACTOR FOR STEPS IN THE PURIFICATION SYSTEM

Fraction	Dry Wt. grams	Amt. R-F[a] nmoles	Recovery[b] %	Purifi- cation
human plasma (ACD, 2.5 liter batch)	200	1725 ±273(6)		
Methanol Extraction				
extract	10.1	1476 ±60(6)	85.6 ±14.0	17 X
residue	--	105 ±37(4)	6.1 ±1.0	--
Sephadex Column				
Fraction I	--	125 ±18(4)	7.3 ±1.6	--
Fractions II + III	1.29	1195 ±129(6)	69.3 ±13.3	108 X
Fraction IV	--	118 ±29(4)	6.8 ±1.2	--
Silicic Acid Column				
major peak	.03[c]	630[d]	36.5	2,440 X
other peaks combined	--	240[d]	13.9	--

[a] mean ± SD(n)
[b] percent of the amount of R-F in the original plasma sample, mean ± SD
[c] estimate based on weighing the contents of two tubes in the center of the major peak
[d] estimate based on area under peaks from graph

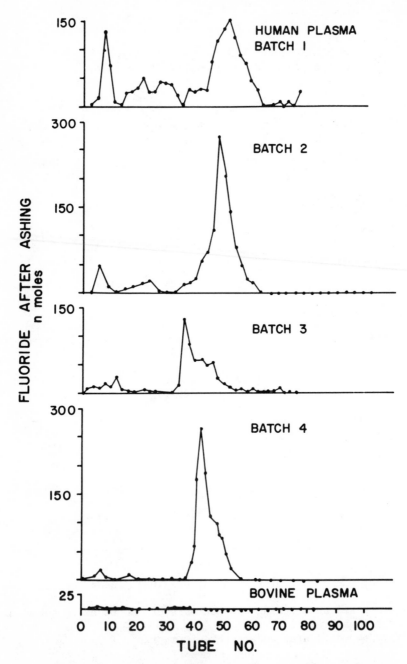

Figure 4. Distribution of organic fluorine from human and bovine plasma in fractions from silicic acid chromatography

which samples were placed.

The blank for the purification process was obtained by using bovine rather than human plasma. No organic fluorine was detectable in the original bovine sample but as a further check the sample was dialysed to remove inorganic fluoride to facilitate making the measurement for organic fluorine by difference. Some organic fluorine was apparent in dialysed bovine plasma: 0.13 ± 0.11 SD μM (n=6), a statistically significant though small difference. This trace amount of organic fluorine clearly was not found in the same silicic fractions as the dominant peak from human plasma as shown in figure 4.

Human plasma had been stored in polyethylene bags with ACD solution. Analysis of ACD solution from unused blood bags and analysis of blood plasma before and after placing it in the bags showed that not more than 5% of the organic fluorine in human plasma could have come from this source.

The distribution of organic fluorine in fractions from silicic acid chromatography are shown in figure 4 for four batches corresponding to 5 liters of the original plasma each. There is clearly one dominant peak lying in approximately the same elution position for each batch (the exact position varied with column use and the degree of hydration of the silicic acid adsorbant). There were always some smaller secondary peaks, but they varied in size and position relative to the major peak.

The sample used for characterization by nmr was obtained by combining the fractions containing the major peaks in each of the four batches. Much of the material from batches one and two had been used for other purposes prior to this combination. The combined sample was rechromatographed on silicic acid and a single sharp peak obtained. The final sample was taken from the central portion of that peak and contained 3.3 μmoles of organic fluorine.

Four sample runs were made on the nmr spectrometer with 15,000 to 17,000 scans each and with a sweep width of 15,151 Hz in all but one run, where it was 7,576. The results of all runs were consistent with the spectrum shown in figure 5 and the chemical shifts shown in table IV. A blank run on the solvent mixture showed no instrumental artifacts which might have contributed to the spectrum. Chemical shifts determined for perfluoro-octanoic acid are also included in table IV. Comparison of the shifts in the unknown with that of perfluoro-octanoic acid show that there is a constant difference in shifts of about 2 ppm except for the $-CF_2-$ peak next to the functional group (peak E) where the shift is about 6 ppm. Only the latter is enough to be considered a significant deviation since external referencing was used for each. The difference in shift for peak E is consistent with the presence of amide or ester derivatives, or possibly with the presence of a sulfonic acid derivative as the functional group. One explanation for the additional peaks in the spectrum is the presence of branched isomers, peaks A and B

Table IV

RESULTS OF NMR SPECTROSCOPIC ANALYSIS

	Chemical Shift[a], ppm		
Peak Designation	Sample	Perfluoro-octanoic Acid	Suggested Assignments
A	-70.7		-CF$_3$ groups at branch points
B	-71.9		terminal -CF$_3$ in branched isomers
C	-80.0		
D	-81.0	-82.6	terminal -CF$_3$ in straight chain
E	-114.3	-120.2	-CF$_2$- next to X[b]
F	-120.3	-123.1	-CF$_2$- in -CF$_2$-CF$_2$-CF$_2$-
G	-121.5	-124.2	
H	-122.3		-CF$_2$- next to branch points
I	-126.0	-127.6	-CF$_2$- next to terminal -CF$_3$

[a]External referencing to CFCl$_3$ was used for the chemical shifts, and these are expressed with positive numbers to lower field (i.e., higher frequency).
[b]where X is likely to be -CO-Y

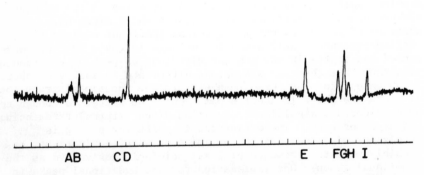

Figure 5.   NMR spectrum of organic fluorocompound(s) isolated from human plasma

representing $-CF_3$ groups at branch points, peak C the $-CF_3$ groups two carbons removed from the branch points, and peak H representing $-CF_2-$ next to the branch points.

The sample was reanalyzed for organic fluorine following characterization by nmr to check for contamination; no additional fluorine was apparent. The degree to which fluorine from perfluoro-octanoic acid is fixed as inorganic fluoride during ashing was found to be 21 ± 3 SD % (n=3).

Discussion

These findings suggest that there is widespread contamination of human tissues with trace amounts of organic fluorocompounds derived from commercial products. All available information on this subject is in accordance with this interpretation. A series of compounds having a structure consistent with that found here for the predominant form of organic fluorine in human plasma is widely used commercially for their potent surfactant properties. For example, they are used as water and oil repellents in the treatment of fabrics and leather. Other uses include the production of waxed paper and the formulation of floor waxes (12). The findings presented here that the concentration of organic fluorine was not related to the concentration of inorganic fluoride either in blood or in the public water supply, and the earlier finding that there was little or no organic fluorine in the blood of animals other than human (6) are all in keeping with environmental sources such as these.

The prevalence of organic fluorine in human plasma is probably quite high since 104 of the 106 plasma samples tested here and all 35 in an earlier study (5) had measurable quantities. The prevalence of the particular compounds isolated and characterized here, i.e., perfluoro fatty acid (C_6-C_8) derivatives, is not known since the starting material for each batch shown in figure 4 was pooled from between 25 and 30 individuals and since only about one third of the original organic fluorine content was accounted for in the fractions containing these compounds (see table III).

Peaks other than the one characterized by nmr appear in the chromatograms shown in figure 4 suggesting that human plasma contains other forms of organic fluorocompounds. They are probably not volatile compounds like freons since it is doubtful that these would be detected by the analytical methods used in this study. They correspond in solubility to very polar lipids since they appear in fractions two and three in the fourth purification step. According to the authors of the method used in that step the first eluent contains most fats, the second and third eluents contain very polar fats like gangliosides and certain bile acids in addition to compounds like urea, phenylalanine and tyrosine. The last fraction contains water soluble non-lipid compounds (11). Components of these other peaks are

less polar than the compounds in the predominant peaks in
accordance with the methanol-in-chloroform gradient used to
elute them in the fifth purification step. Other forms not seen
in silicic acid fractions may also exist since only about half
the original organic fluorine was recovered in these fractions.

The actual amounts of the perfluorinated fatty acid
derivatives in human plasma is not known both because
individual plasma samples were not assayed for these particular
compounds and because the degree to which organic fluorine from
these compounds is converted to inorganic fluoride during ashing
is not known. Metal salts of perfluorinated fatty acids have been
reported to decompose at 175 to 250° C forming CO_2, volatile
perfluorinated olefins one carbon shorter, and one atom of
fluoride per molecule (13). About 3 fluorine atoms per molecule
of perfluoro-octanoic acid were fixed as inorganic fluoride by
ashing methods used here. Thus, values reported here for
fluoride after ashing fractions from the major peaks in figure
4 probably represent somewhere between one-third and one times
the molar amount.

Little has been published about the metabolic handling and
toxicology of perfluorinated fatty acid derivatives. Computer
assisted literature searches using Medline, Toxline and Chemcon
developed no information on these subjects. This was surprising
with respect to the widespread commercial use of such compounds.
It would appear from information presented here that rapid
excretion of such compounds into urine is unlikely since they
are bound to albumin in the blood. On this topic it can also be
stated that other chemicals are usually not toxic in blood concentrations similar to those found here for organic fluorine.

The concentration of organic fluorine in human plasma may
be changing with time. In 1960 Singer and Armstrong reported
that the plasma of 70 individuals residing in communities with
1 ppm or less fluoride in their public water supply had an
average concentration of fluoride of 8.8 μM (8). They prepared
their samples by ashing them and then distilling fluoride from
the ash acidified with perchloric acid (7). Thus, it seems
likely that their values for "fluoride" would have included
organic fluorine had it been present. Assuming that inorganic
fluoride concentrations at that time were similar to those found
in this study (see table II), the organic fluorine component
would exceed 7 μM. In 1969 the same investigators using the
same method reported an average fluoride concentration of 4.5 μM
for 6 plasma samples each pooled from at least 3 individuals
supposedly living in fluoridated communities (14). This
corresponds to an organic fluorine component of only about 4 μM.
Organic fluorine concentration presented here averages only
1.35 μM. Therefore, there may have been a decrease in the
concentration of organic fluorine in human plasma since the late
1950's. An alternate explanation might be that differences in

the analytical methods or differences in the sample populations caused these values to vary.

Organic fluorine is the predominant form of fluorine in human blood except where the concentration of fluoride in drinking water is high (in which case fluoride predominates, see table II). This together with the finding reported here that there is no apparent relationship between the concentrations of organic fluorine and inorganic fluoride in plasma helps explain why in earlier studies (8) no relationship was found between plasma fluoride determined in ashed samples and the fluoride content of the public water supply. The data in table II show that when methods specific for inorganic fluoride are applied, a clear relationship between fluoride in plasma and fluoride in the public water supply (between 0.1 and 5.6 ppm) can be demonstrated. Thus, there is no need to postulate the existence of such rigorous homeostatic control mechanisms for plasma fluoride as suggested earlier (8). Average plasma fluoride concentrations for individuals living in the same city as reported here reflect the balance established between fluoride in blood and that in bone mineral over periods of years. These findings do not contradict a passive homeostatic control mechanism in which bone mineral damps swings in blood fluoride concentration over relatively shorter periods of time.

The values presented here for the average inorganic fluoride concentration of plasma from individuals living in a community having about 1 ppm fluoride in the water supply are consistent with recent findings of others using similar methods (14, 15).

Literature Cited

1. Taves, D.R., Nature (1966), 211, 192.
2. Taves, D.R., Nature (1968), 217, 1050.
3. Taves, D.R., Nature (1968), 200, 582.
4. Taves, D.R., Talanta (1968), 15, 1015.
5. Guy, W.S., "Fluorocompounds of Human Plasma: Analysis, Prevalence, Purification and Characterization", doctoral thesis, University of Rochester, Rochester, N.Y., 1972.
6. Taves, D.R., J. Dental Res. (1971), 50, 783.
7. Singer, L. and Armstrong, W.D., Anal. Chem. (1959), 31, 105.
8. Singer, L. and Armstrong, W.D., J. Appl. Physiol. (1960), 15, 508.
9. Taves, D.R., Talanta (1968), 15, 969.
10. Frant, M.S. and Ross, Jr., J.W., Science (1966), 154, 1553.
11. Siakotos, A.N. and Rouser, G., J. Amer. Oil Chem. Soc. (1965) 42, 913.
12. Bryce, H.G., "Industrial and Utilitarian Aspects of Fluorine Chemistry" in "Fluorine Chemistry", Vol. V, J.H. Simon, ed., Academic Press, N.Y., 1964.
13. Hals, L.J., Reid, T.S. and Smith, G.H., J. Amer. Chem. Soc. (1951), 73, 4054.
14. Singer, L. and Armstrong, W.D., Arch. Oral Bio.(1969),14,1343.
15. Singer, L. and Armstrong, W.D., Biochem. Med. (1973), 8, 415.

Q. I wonder if you tried to correlate within individuals the level of organic fluorine with age.

A. It would certainly be interesting to have this information but unfortunately we cannot supply it at this time. An expeditious approach might be to analyze cord blood from infants of mothers who had not received fluorine-containing anesthetics at childbirth. It would also be of interest to know whether individuals living in isolated regions have organic fluorine in their blood plasma.

Q. Did you say the sample analyzed by nmr contained methyl alcohol?

A. Yes, I did.

Q. Methyl alcohol will react very rapidly with fluorinated acids. The nmr spectrum may, therefore, represent that of methyl ester derivatives.

A. Methanol was also used in the last three steps of the purification system. The nmr spectrum is consistent with the presence of methyl ester derivatives of perfluorinated fatty acids (C_6-C_8) and their branched isomers.

Intravenous Infusion of Cis-Trans Perfluorodecalin Emulsions in the Rhesus Monkey

LELAND C. CLARK, JR., EUGENE P. WESSELER, SAMUEL KAPLAN, and CAROLYN EMORY
Children's Hospital Research Foundation, Elland and Bethesda Aves., Cincinnati, Ohio 45229

ROBERT MOORE
Sun Ventures, Inc., Marcus-Hook, Pa. 19061

DONALD DENSON
Stanford Research Institute, Menlo Park, Calif. 94025

Introduction

Interest in the possibility of making fluorocarbon-based artificial blood began following the discovery by Clark (1) that animals survived the breathing of oxygen-saturated FC75 (3M Company). Since that time over 150 papers have been published, a FASEB Symposium (2) and several industrial symposia have been held on the subject. Reviews by Sloviter (3) and Geyer (4) have been published. A large number of perfluorochemicals (PFC) have been screened for this purpose. Most PFC carry oxygen in biologically significant quantities. But most are ruled out for practical use because, after performing their oxygen-carrying function, they remain in the body, largely in the liver, after being taken up mainly by the mononuclear phagocytes as are all PFC. So far it seems that only those containing fluorine and carbon or fluorine, carbon and bromine in their structure leave the body. Although we have not completed our testing of the PFC on hand and expect to test new ones in the future, nonetheless, of the PFC tested so far, perfluorodecalin (PFD) emerges as the best because it leaves the liver in a reasonable time and has a vapor pressure compatible with intravenous use. As an emulsion, its acute intravenous toxicity in the mouse is very low. It is available in commercial quantities and can be partially purified by distillation. For these, and other reasons, it was selected for testing as artificial blood in a non-human primate, the rhesus monkey. We previously reported (5) on the infusion of perfluorodecalin in one awake rhesus monkey. This paper describes the first results of further tests in 19 monkeys and 10 dogs and outlines the beginning of a protocol for its possible evaluation as a blood substitute in man. Of the 19 monkeys, 10 were infused with 10% PFD, 4 with 20% PFD emulsions, and 4 with only a "hypotensive test dose".
The data included here is only a part of that on hand; it has been selected to illustrate the salient problems and findings. It is to be interpreted in terms of our previous publications (6-18) on this subject.

Data on oxygen solubility in PFC presented at this meeting
is being expanded and will be the subject of a separate report
(19).

Methods and materials

Nineteen monkeys were selected from the Institute of Develop-
mental Research's resident population. Many of these monkeys had
been previously used for testing drugs in teratological research.
Three were males and had not been given drugs. The monkeys were
housed in stainless steel cages in air conditioned rooms and were
fed a diet of Purina monkey chow (Code 5038) supplemented with
bananas, oranges, and raw peanuts. They were given periodic skin
tests for tuberculosis and were given 11 mg/kg isoniazide daily
impregnated as a solution on a sugar cube. The monkeys, all of
which presumably have lung mites, were chest x-rayed on arrival
and weighed weekly. Public Health Certification Procedures were
used for monkeys obtained from India and Helsinki rules were followed in the laboratory. The dogs were beagles obtained from a
commercial source.

The first monkey tested and reported (5), was a young male.
Two of the monkeys in the present series were also males. The
veterinarians were unable to assign an approximate age to any monkey but, judging by their teeth and their behavior, many were very
old.

The perfluorodecalin used in these experiments was purchased
from ISC Chemicals, Ltd., Avonmouth, Bristol, England and purified
by spinning band distillation using a Perkin-Elmer NFA-200 Auto-
annular still. The fraction boiling between 91°C and 93°C at
180 mm pressure was used for these studies.

The Pluronic (PF68) solution was made by dissolving 200 gm in
sterile water (Abbott), brought up to 1 liter, and filtering suc-
cessively through 5.0, 0.8, and 0.22 µM Millipore filters at 4°C.
This stock was diluted just before use, the ionic composition was
adjusted so the emulsion contained half the salts required for
Ringer's and the pH was adjusted with Tham (28).

Emulsions were made in a Gaulin Model 15M lab homogenizer.
Parts which were exposed to the emulsion were autoclaved. The
shear pressure was preset to 6000 lbs/in^2 and the gauge was removed. Homogenization was continued until the optical density
plateaued. The emulsion was placed in sterile Pyrex vials or one
liter bottles, frozen and preserved at -70°C.

Sterility tests, performed by adding 2 ml into 20 ml of a
supplemented peptone broth prepared by Becton-Dickinson and Co.,
were conducted on random samples of the emulsion. All tests showed the emulsion to have no bacterial growth.

Chemical methods of sterilization including the use of pro-
piolactone were avoided, except with one monkey, Melvin (5), who
received emulsion sterilized with this compound.

Two batches of 10% by volume PFD in 5% PF68 and two batches of 20% by volume PFD in 10% PF68 were prepared. All of the blood pressure test doses were from the same batch (1 liter) of 10% PFDE and were taken from small Pyrex vials which were thawed just before use. All of the monkeys infused with 10% PFDE received emulsion (4 liters) from the second batch where individual bottles were thawed just before use. Two of the monkeys, C2 and C3, received 20% PFDE from a third batch (1.5 liter). Two, 74 and 113, received 20% PFDE from a fourth batch (4 liter). The first three batches were prepared from pooled PFD (ISC analytical reference S173/A) while the fourth batch was prepared from PFD (ISC analytical references S173/A and S148/B). All the 10% PFDE and the first two 20% PFDE emulsions were stored in ice water until used but the 20% PFDE for 74 was warmed before administration and that used for 113 was bubbled with oxygen at room temperature for three hours before infusion.

Blood samples were collected from the anesthetized monkeys through indwelling catheters and cannulae using sterile plastic syringes having the dead space filled with heparin (1000 USP units/ml). The samples from awake restrained monkeys were obtained by venipuncture. Precautions were taken to be sure that the blood samples were not diluted with the isotonic solutions used to rinse the sampling lines. Samples for blood gases, pH, and packed cell volume were analyzed immediately. The remaining blood was centrifuged and the plasma analyzed by Autotechnicon SMA 12 procedures (21). Venous blood samples were taken at random from L. Clark, to determine the variability of the same blood factors being measured in the monkey. Most of the samples from L. Clark were taken during the fasting state, all those reported here were taken while L. Clark was in a state of well-being, and all were processed like monkey blood. Samples were taken periodically from the monkey previously published (5). All samples not analyzed immediately were chilled in an ice-water bath. Approximately 4800 determinations were performed as part of the research reported here.

Test for hypotensive effect

The monkeys were fasted overnight and anesthetized with intravenous sodium pentobarbital in the morning. They breathed humidified oxygen. The arterial pressure was measured by means of a pediatric cuff and a Model 802 Doppler (Parks Electronics Laboratory, Oregon) using the radial artery. 0.05 ml/kg of emulsion was injected intravenously and rinsed in with 5 ml of Ringer's solution. A second test dose was given within five minutes after the first in the monkey but after the blood pressure returned to normal, 10 or 20 minutes, in the dog. A similar procedure was used for each vial of emulsion using the beagle within one day of the test in the monkey. The blood pressure in the dog was measured with a mercury manometer and direct cannulation of the femoral artery.

Twelve monkeys were subjected to biopsy before PFDE infusion and nine after. The overnight-fasted monkeys were anesthetized with sodium pentobarbital and a 2 or 3 gram piece of liver was excised under direct vision, fixed, and examined by light and electron microscopy using methods previously published (18). Parts of these post infusion samples were analyzed for PFD by sodium biphenyl combustion and by GLC. Control blood samples were taken prior to making the biopsy incision. Post biopsy blood samples were taken one hour after the incision was closed and again on the following day from the restrained animal.

For the infusion, the animal is anesthetized with pentobarbital and kept asleep with pentothal. An endotracheal tube is inserted and the animal's head covered with a transparent bag being flushed with humidified oxygen. The femoral veins and the radial artery are cannulated with sterile plastic tubes and the right heart is catheterized with a Berman angiographic balloon catheter.

On the day of infusion when the blood pressure, respiration, and ECG of the monkey have been recorded for several minutes and look stable, the first samples of blood are taken for blood gas analyses. A 2 ml sample of blood is drawn into a heparinized 2 ml syringe from the right heart and the radial artery. If the pO_2 and pCO_2 readings obtained on these samples are within normal limits, a large (12.5 ml/kg) sample of blood is removed. This blood is centrifuged and analyzed as described elsewhere. At this point 21 ml/kg of 5% human albumin is infused with a Sage Instrument syringe pump at 6 ml/min.

Another 12.5 ml/kg of blood is removed carefully so that the blood pressure does not fall below 80 mm Hg. The hematocrit of this blood is determined and if it is not one half the original hematocrit, another 4 ml/kg is infused and another hematocrit run. To date, this procedure has always reduced the hematocrit approximately 50%. In five of the first six monkeys infused 50 ml/kg of Ringer's solution was given in place of the albumin.

The PFDE is infused in four batches and arterial and mixed venous blood gases and pH are measured after each of the four batches are in.

After the infusion, the arterial cannula and all but one of the venous cannulae are removed and the incisions closed. The animal is kept in the operating room until it begins to waken. Then it is transferred to a recovery room and closely watched for the next 12 hours. Fluids may be given intravenously if required. Usually the monkeys are awake and drinking liquids or even eating by late afternoon.

The autopsies are conducted under the auspices of the Medical School's Veterinary Department. The organs are inspected, photographed, weighed and half of each organ is placed in neutral 10% phosphate-buffered formalin solution and half is placed in 95% ethanol after portions of each organ are removed for fixing, staining and examining under light microscopy.

Results and discussion

Interlaboratory variation in the analysis of three perfluorinated liquids is given in Table 1. It is apparent that the difficulty lies in getting quantitative recovery of fluorine and not
of carbon. Some analytical laboratories obtained such inaccurate
results that they abandoned attempts to analyze the samples. Another laboratory is continuing its efforts to develop a method.
No difficulty was encountered, in preparing the PFD emulsions. Those preserved at -70°C remained for months with no discernible change. The seven day LD50 in mice of the infused 10%
PFDE was 140 ml/kg; the LD50 of the 20% PFDE was 69 ml/kg.
Measurement of particle size distribution in three emulsions
was performed at Sun Ventures by a technique (26) involving electron microscopy and the results are shown in Figure 1.
Tables 2 and 3 represent the changes in blood chemistry found
upon biopsy of the liver. The changes in glucose probably have
little meaning because special efforts were not made to prevent
glycolysis. The increased lactic dehydrogenase activity may have
come from the damaged liver. There is no doubt that creatine
kinase increased as a result of the procedure but it is known to
increase after any kind of surgery (25).
Control values for blood samples from a human primate are
shown in Table 7. and those from awake monkeys in Table 5.
One reason the monkey was selected for these experiments is
that it very often behaves like the human in its response to drugs.
The dog sometimes shows bizarre responses to emulsions and solutions containing surfactants. In Table 4 it can be seen that no
monkey tested, but all dogs, suffered a drop in blood pressure.
It sometimes takes half an hour for the dog to recover but once it
has recovered a second dose has very little or no effect.
In Tables 10 and 11 it can be seen that neither the anesthesia nor the surgical manipulations involved in giving a test dose
had a significant effect upon the blood components analyzed except for creatine kinase where a small but significant increase
occurred. The abnormally high values for twenty-four hours on
monkey 80 are due to the fact that she was moribund and died two
hours later after several attempts at resuscitation.
The details concerning the replacement of withdrawn blood by
Ringer's solution and/or 5% human albumin are given in Table 16.
The amounts of PFDE infused are shown. For our purpose here we
have referred to Dextran 40 and albumin as osmotic or oncotic liquids. Albumin, however, has a longer half life in the body and
of course the PF68 in PFDE has oncotic activity. 5% human albumin was given previously to one monkey with no discernible effect.
That we had considerable difficulty in maintaining blood balance after phlebotomy and PFDE infusion is apparent from the table.
Most of the difficulties happened during the evening of the day of
the infusion and were thought to be due to the fact that most of
the PF68 and water administered had been excreted leaving the

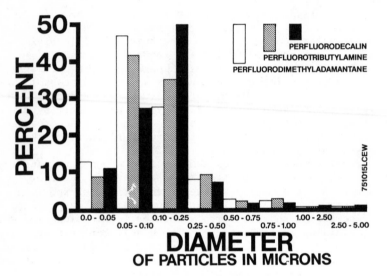

Figure 1.   Particle size distribution as determined by electron microscopy

animals hypovolemic and although awake they were not alert and were weak. When Dextran 40 was given on two occasions (C2, C3) four or five hours post infusion the animals responded to the point where they were difficult to restrain. They also drank Gatorade and ate. One which received Dextran twenty-four hours later, following an all night slow bleeding from a poorly tied in cannula, revived only briefly. Most of the 10% PFD monkeys that received albumin during the phlebotomy required less fluid later. Two (74, 113) that received 20% PFDE died, possibly because they were overloaded, but more likely because the PFDE they were given had been at room temperature too long and coalescence of particles had begun. Our experience in maintaining blood and fluid balance in thousands of dogs and human patients has not served us well with the rhesus. It seems to us that the rhesus monkey, particularly the old monkey, is a very delicate and fragile animal.

The main finding in Table 6 is that the mixed venous oxygen tension increased in <u>all</u> the animals receiving PFD emulsion. In none of those receiving 10% PFDE did the mixed venous pO_2 increase beyond 100 mm, the approximate point where all the hemoglobin is saturated. In three of the monkeys receiving 20% PFDE the mixed venous pO_2 went above 100 mm. In monkey C3 the arterial pO_2 was on the low side and this could not be increased by chest and/or heart massage. At autopsy this monkey was found to have severe emphysema. This low arterial pO_2 may account, at least in part, for the low venous pO_2. At autopsy C3 was found to have lungs heavily infested with mites.

The heart rate decreased and the respiration increased as a result of PFDE infusion as shown in Table 8. The T tests shown in this table indicate there is no difference in these responses to the 10% and 20% PFDE.

There was no change in arterial pressure as a result of PFDE infusion as shown in Table 9. From these and other studies we have concluded that there is an increase in pulmonary arterial pressure followed by a return to normal in about an hour. Because of uncertainty about the location (RV or PA) of the catheter tip (x-ray equipment was not available to us during these studies) the data in Table 8 cannot be analyzed statistically. However it can be seen that in all of the cases where the pressure was low before infusion, probably indicating that pulmonary arterial pressure was being monitored, there was a distinct increase.

Tables 12 and 13 give the results of analyzing the blood of monkeys receiving 10% PFDE and Table 13 gives the average values and standard errors for this data. Some of the apparent decreases in concentration of blood components are due to the fact that about half the blood was removed and diluted by Ringer's, albumin, and PFDE. The extent to which dilution <u>per se</u> affected the results can be best judged by the extent to which the hematocrit decreased. While the table shows a decrease in all components except total bilirubin, if these are "corrected" for dilution by

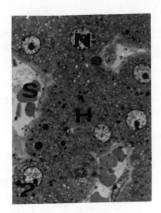

Figure 2. Light microscopic view of the liver of a monkey before infusion. H, hepatocyte; N, nucleus; S, sinusoid; Toluidine blue 0. × 850.

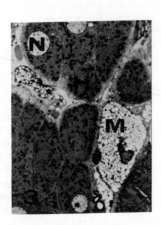

Figure 3. Liver specimen taken at sacrifice two weeks after infusion with a 20% emulsion shows hepatocytes essentially unchanged morphologically. Cytoplasm of mononuclear phagocytes (Kupffer cells and macrophages) is completely filled with particles of PP5, and bulges into the lumen of sinusoids. H, hepatocyte; N, nucleus; M, mononuclear phagocyte. Toluidine blue 0. × 850.

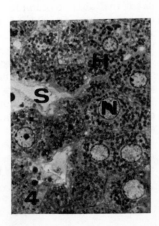

Figure 4. Eleven months after infusion, the liver shows normal morphology and no structures remained which could be identified unequivocally as fluorocarbon. Other features were within normal limits. H, hepatocyte; N, nucleus; S, sinusoid. Toluidine blue 0. × 850.

dividing by 0.32, the hematocrit factor, the results indicate an increase in everything except cholesterol. The percentage change in the components before and after correction are as follows: TP 41,128; AB 71,222; CA 71,222; IP 90,281; GL 74,231; BN 88,275; VA 44,138; CT 80,250; TB 140,438; AP 51,159; LD 75,234; GO 45,141; CK 142,444; CH 18,56; PV 32,100.

The blood chemistry changes found before and after infusion of 20% PFDE are shown in Table 14.

Twenty-eight samples were taken from the monkey infused with 10% PFDE about a year ago and the mean values are shown in Table 15. The mean alkaline phosphatase is higher than that shown in Table 3 and Table 5 but this would be expected because Melvin is growing rapidly. All the other data are normal according to the information we have accumulated here.

Over a three month period the average weight gain of the eight surviving monkeys infused with PFDE was not significantly different from six monkeys selected at random as controls.

Morphology. Three of nineteen monkeys are pictured here which represent the preinfusion, post infusion and long term recovery of the liver of all monkeys examined to date.

The normal, or preinfusion morphology of the liver differed slightly from that of other commonly used laboratory mammals (fig. 2). Variations included a marked elevation in hemosiderin deposits in Kupffer cells, generally associated with age, numerous large autophagic vacuoles (10 μ) and the presence of long tubular structures in the mitochondria. These peculiarities have also been observed by others (27).

Within twenty four hours after infusion of emulsions of PP5, particles begin to appear within the phagocytic cells of the sinusoids of the liver, other organs, and the circulation. Just after infusion, while particles are in the bloodstream, the polymorphonuclear cells show a few cytoplasmic particles. Most affected morphologically by the infusion of emulsions are the cells of the mononuclear phagocytic system. These cells engulf individual particles or clumps of particles until they reach very large proportions (fig. 3) as they rest in the liver sinusoids, splenic pulp or similar area. After a few days they may aggregate into small nodules of epithelial-type cells similar to those seen in other foreign body responses (fig. 3).

The hepatocytes are not changed dramatically by the infusion of emulsions (fig. 4) but a small number of particles do enter the cytoplasm. These diminish, however, and ultimately disappear by unknown means. Residual effects of the occupation of the liver (and of course other comparable cell systems in the body) by PFC in fact have not been observed morphologically, since mitochondria, smooth and rough endoplasmic reticulum, lysosomes, microbodies, nuclei, etc., appear indistinguishable from controls (figs. 5,6). All the specimens shown were obtained at biopsy.

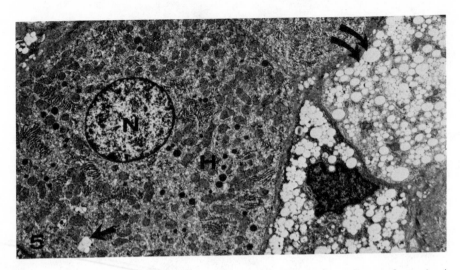

Figure 5. Electron microscopic view of liver from the same monkey as figure 3. A very small percent of the cytoplasm is occupied by particles of fluorocarbon (small arrow). Organelles appear unchanged. H, hepatocyte cytoplasm; N, nucleus; M, mononuclear phagocyte with cytoplasm full of fluorocarbon particles (large arrows). Uranyl acetate, lead citrate. × 8,500.

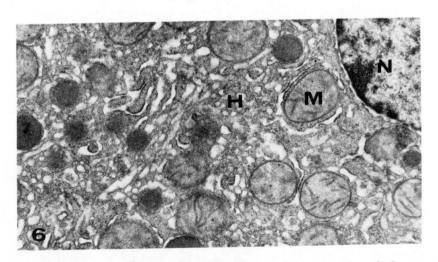

Figure 6. Liver from the same monkey as figure 4 shows organelles which appear normal. Fluorocarbon particles, no longer identified in the cytoplasm of hepatocytes or mononuclear phagocytes, have apparently left the liver unchanged. M, mitochondria; N, nucleus; H, hepatocyte cytoplasm. Uranyl acetate, lead citrate. × 4,000.

Post mortem findings. Of ten monkeys which received 10% PFDE, eight are alive. One, Melvin, has survived over a year. One died the day after infusion from slow bleeding around a venous cannula. One (117) died eleven weeks after the infusion from anesthesia preparatory to surgery for liver biopsy. Of the four monkeys which received 20% PFDE two died within twenty four hours following the infusion and were thought to have died because the emulsion had begun to deteriorate. Possibly their circulations were overloaded. Two survived in apparent good health but were sacrificed at one and two weeks post infusion because their legs became ischemic and possibly gangrenous, following ligation of their femoral arteries as a part of the process of recording their arterial pressure. Femoral arteries can be ligated in cats and dogs with no visible effect but primates are susceptible to necrosis if the femoral arterial circulation is compromised. One monkey's femoral artery was successfully repaired; this animal (117) lived eleven weeks with normal legs. Following the problem with the femoral artery, direct pressures were measured from the radial artery. One monkey suffered partial loss of the fingers of the left hand following ligation of the radial. One injured monkey, 6, was sacrificed shortly after it was purchased, and before it was used as a control for organ weights. Monkey 80 bled to death overnight because a femoral artery was punctured during an attempt at catheterization.

The organ weights obtained at autopsy are shown in Table 17. Because of the dearth of normal data in the rhesus very little can be said about the effects of infusion. The liver of C2 and the spleen of C3 are quite possibly enlarged.

All the monkeys were found to have lung mites, and some were considered heavily infested. Many of the lungs were stiff and did not collapse.

More detailed reports of the findings with light microscopy will be the subject of a separate report.

Sixteen pieces from various parts of the liver of one monkey were analyzed and were found to range from 0.61% to 2.05% with a mean of 1.45%.

Information obtained by sodium biphenyl combustion of the liver and GLC are shown in Table 18. There was of course circulating PFDE in monkeys 62, 74, and 113. Depending on the dose, it requires several days for it to completely disappear from the blood. The PFD disappears from the blood both by evaporation through the lungs and skin and by being engulfed by the scavenger cells of the body.

Interestingly, a trace could be detected in Melvin's liver almost a year after the infusion.

There is considerable variability between monkeys in the way in which they sequester the PFD in their livers and spleen, and the rate at which it leaves. The half life of PFD in the liver is probably about two weeks. In the mouse, most PFD is gone in two weeks.

General discussion

Although most of our results have been discussed above we think it may be desirable to add some general comments about our concepts and our findings.

Our work on artificial blood is prompted by our experience with thousands of open heart surgery patients and a familarity with the problems surrounding plasma, plasma substitutes, parenteral solutions and the use of donor blood.

Various substances have been used over the past several decades as plasma substitutes. Certain of these substances, such as Dextran and plasma albumin, have earned a definite place in clinical medicine for maintaining circulatory volume. Generally, such solutions dissolve only 2 or 3 ml of oxygen per 100 ml, while whole blood dissolves about 20 ml per 100 ml.

In order to meet the oxygen demands of the body the cardiac output must be increased if the red cell mass decreases. In other words, the way in which the body attempts to compensate for oxygen carrying capacity of blood is to work harder and pump more blood. Because fluorocarbons carry large quantities of oxygen they can be used to increase the oxygen capacity of the circulating blood and therefore decrease the work of the heart. Of course, a weak, failing, or injured heart may not be able to increase its output to meet oxygen needs and shock and death may follow. Therefore, a primary function of artificial blood is to decrease the work of the heart. Oxygen carrying liquids will be most useful in conditions where the cardiac output is low or the blood volume is below normal.

The only other way to increase the oxygen capacity of blood is to add stroma-free hemoglobin but this has not proven to be practicable yet. Hemoglobin is easily stored and it may someday be possible to use hemoglobin from other animals, such as the cow and the pig. One of its faults is its high rate of excretion. Synthetic oxygen chelates seem to be far in the future. PFC artificial blood, like stroma-free hemoglobin, has no blood types.

Our research on liquid breathing of perfluorochemical liquids formed the early basis of the work reported here. It indicated, for example, that certain highly fluorinated liquids were probably biologically inert. It has led to the use by Dr. David M. Long, as we hear today, of brominated perfluorinated liquid as X-ray contrast agents in the diagnosis of lung disease. Perfluorocarbon liquids may also be useful some day in the treatment of lung diseases, such as by washing out obstructive materials.

It would be highly desirable to have a suitable water soluble synthetic oxygen solvent to increase the oxygen capacity of plasma or plasma substitutes. PFC in general are good oxygen and carbon dioxide solvents but they are completely insoluble in water. Therefore they can only be used as emulsions. This introduces at least three problems. (a) Biologically suitable emulsifiers must be found and used. (b) Certain PFC cannot be emulsified and certain others form only unstable emulsions.   (c) Because emulsions consist of suspended particles they are gradually removed from the circulation and deposited in part in the liver and spleen.

We have elected to concentrate our efforts on a study of the use of perfluorodecalin primarily because it gradually leaves the body (16), but also because it has an acceptable vapor pressure, and is a good oxygen and carbon dioxide solvent. It readily forms an emulsion having a low optical density but must, at present, be preserved at -70°C. For comparison, we and others have found that FC47 forms a fine particle emulsion, as judged by optical density, which is far more stable than that produced by PFD but once entrapped in the liver it remains for years. PFD in addition, unlike, for example, perfluoromethyldecalin with ten isomers, has only two.

A fluorocarbon having but a single molecular configuration, such as adamantane would be preferable to having one with two isomers such as PFD. PFD after purification by spinning band distillation as described in this paper, was analyzed at Stanford Research Institute by GLC and field ionization mass spectrometry and reported to be 96.4% pure with possibly eight distinct compounds present having molecular weights ranging from 68 to 484.

A few milliliters of the two isomers of PFD, having the published properties, were laboriously prepared by gas chromatography. It may be that the very property of being inert, which makes some of these fluorocarbons potentially useful in medicine, makes them difficult to purify, characterize, and identify. Of course, high purity and uniformity are required for medical use.

The list of useful emulsifiers is very short indeed because they must be non-toxic, non-hemolytic and cause no undesirable physiological reactions. Only two such substances exist at the present time. Special egg phospholipid (Vitrum, Sweden) has been used extensively outside of the United States as an emulsifier for intravenous fat emulsions for clinical use. This emulsifier can only be used to formulate emulsions under very carefully controlled conditions, as to preparation, temperature, exposure to air and other factors. It is capable of emulsifying PFD. The other emulsifier, Pluronic F68 (Wyandotte, USA), at the present time is not used clinically in the USA as a component of intravenous emulsions of any kind. It has found extensive use as an emulsifier for PFC emulsions in research with animals. It was selected for the PFDE used here in the rhesus because it forms emulsions with PFD and is relatively non-toxic and stable.

Most of previous work with PFDE was done using the mouse, the cat, and the dog. We decided to perform the tests reported here in the rhesus (macaca mulatta) monkey to determine if the primate responded differently.

We found that the hypotensive effect of small doses of emulsion did not occur in the primate. Measurements of the LD50 of plasma from the dog at the depth of the blood pressure drop

indicated that no toxic substance was present and we tend therefore to agree with Wretlind (23, 24) that this is some kind of a
cardiovascular reflex.

Emulsions prepared from PFD and PF68 are difficult to characterize as to particle size and concentration of PF68 in the aqueous phase after emulsification. There is no one method for measuring particle size in the ranges involved here which is generally accepted. Optical density measurement continues to be the best
means to monitor the making of emulsions from any given PFC. It
probably cannot be used to compare particle size when different
PFC structures are involved.

Space will not permit discussion of the complex factors involved in maintaining fluid balance by means of osmotic, oncotic,
and hydrostatic forces. Suffice to say that we attempt to maintain oncotic activity by the short-acting Pluronic and Dextran and
by the long acting plasma albumin. We found that Dextran given
several hours after the infusion of 20% PFDE had a dramatic bene-
ficial effect; it has brought animals from a near unconscious
state to one of near normal.

The difference in the fate of the two monkeys given a warmed
20% emulsion and a cooled 20% emulsion would seem to suggest that
warming an emulsion in vitro is apparently completely different
than warming it in vivo. Not only does the emulsion appear to be
much more stable in vivo than in vitro, even though far above room
temperature, but judging from the bluish haze in the plasma of
some of these monkeys, the particle size may even have decreased.

This may be due not only to the emulsifying properties of
blood, but to the mechanical mixing effects in the cardiovascular
system and the unique characteristics of the lining of the blood
vessels.

It should be borne in mind that most of the chemical analyses
of blood reported here are done on plasma dialyzed through a cel-
lophane membrane and should therefore not be affected by the presence of PFD. Total protein, albumin, total bilirubin, and lactic
dehydrogenase were analyzed without passing through a dialysis
membrane and the measurement could have been affected.

Summary

Of nineteen monkeys selected for this study, fourteen were
used for the infusion of purified perfluorodecalin in the form of
10 and 20% by volume emulsions. Four received only test doses and
one was sacrificed before use as a control. Pluronic F68 was used
as the emulsifying agent. Previous to the infusion of PFDE, blood
was removed to decrease the hematocrit by half and this blood was
replaced by either Ringer's solution or 5% human albumin. Nine
out of ten monkeys infused with 10% PFDE survived. Two of four
monkeys infused with 20% PFDE survived. Mixed venous pO_2 increased in all monkeys and exceeded 100 torr in three monkeys, which
received 20% of PFDE. The surviving monkeys appear to be in good

health. Three monkeys were sacrificed, one as a control, two because of compromised circulation to the leg. One succumbed during
anesthesia preparatory to biopsy and one from a surgical accident
during cannulation. Aside from a drop in cholesterol there were
only questionable changes in about 20 blood components analyzed.
Small (0.05 ml/kg) doses of PFDE given to dogs invariably caused a
pronounced drop in blood pressure but none occurred in the monkeys. The morphologic changes in the liver of the monkeys were
reversible with no sign of damage. The major problems encountered
were difficulty in obtaining pure PFD, making stable emulsions,
obtaining analysis of compounds and tissues for PFD, working with
the fragile cardiovascular and respiratory systems of the rhesus
and judging optimum fluid balance post phlebotomy and post infusion.

Abbreviations

NO	Number
SP	Samples
TP	Total Protein (gm%)
AB	Albumin (gm%)
CA	Calcium (mg%)
IP	Inorganic Phosphrous (mg%)
GL	Glucose (mg%)
BN	Blood Urea Nitrogen (mg%)
UA	Uric Acid (mg%)
CT	Creatinine (mg%)
TB	Total Bilirubin (mg%)
AP	Alkaline Phosphatase (mU/ml) (EC 3.1.3.1)
LD	Lactic Dehydrogenase (mU/ml) (EC 1.1.1.27)
GO	Glutamic-oxaloacetic Transaminase (mU/ml) (EC 2.6.1.1)
CK	Creatine Kinase (mU/ml) (EC 2.7.3.2)
CH	Cholesterol (mg%)
PV	Packed Cell Volume (%), hematocrit
PF	Packed PFD Volume (%), fluorocrit
PFC	Perfluorochemical
PFD	Distilled Perfluorodecalin
PFDE	Perfluorodecalin Emulsion
C	Control
GLC	Gas-Liquid Chromatography
MK	Monkey
PTD	Post Test Dose
M	Mean
SE	Standard Error = $S.D./\sqrt{N-1}$
T	Fisher's t-test of significance
RI	Ringer's solution
BL	Blood
LD50	Mean lethal dose, i.v. injection
ME	The monkey (ref. 5), named Melvin
PP5	In the section on morphology this is used to designate dis- tilled perfluorodecalin.

Acknowledgements

The authors wish to thank Dr. George Miller for guidance in surgical techniques used for tissue biopsy. We consulted Christ Tamborski concerning chemical problems. The assistance of the following persons is gratefully acknowledged: Dr. Marian L. Miller, Dr. Steele F. Mattingly, Dr. Jag Lal, Frank Knapke, Pat Turner, Lilam Stanley, Margaret Kelm, Steven Jones, David DeForest, Barbara Cincush, Stanley Gaines, Eleanor Clark, Eleanor Brinkmoeller, Sandra Hoffman, and Dotty O'Reilly of Sun Ventures. The Pluronic F68 was a gift from Dr. Irving Schmolka of Wyandotte. This research is supported in part by grants HL17586, HL17353, GM21475, HD05221, from the National Institutes of Health, grant 74 619 from the American Heart Association and a grant from the Southwestern Ohio Chapter of the American Heart Association.

Shortly after mailing our manuscript we received, from Technicon Industrial Systems, Tarrytown, New York 10591, via Richard Carr, 1000 Crest Circle, Cincinnati, Ohio 45208, additional data for normal values for clinical blood chemistry in the rhesus. One report (29) gives the normal values for blood from forty-six female and thirty-three male rhesus. Another report (30) gives the mean and range for eleven blood component analyses for fifty rhesus.

Trade Name	PP5		Fomblin		E4	
Structure	(perfluorodecalin ring structure)		CF_3 $[O-CF-CF_2]_n$ $n \cong 6$		$CF_3CHF(OCF_2CF)_4F$ CF_3	
	%F	%C	%F	%C	%F	%C
Calculated	74.01	25.99	67.70	21.40	70.28	21.43
Lab A	74.19	26.10	72.65	21.14	72.93	21.90
	73.74	26.35	72.73	21.32	73.39	22.01
	73.53		71.73			
	73.94		72.32			
			71.85			
Lab B	78.24		84.05			
	78.32		83.90			
Lab C	66.35		65.76			
	77.02		73.59			
	76.72		73.79			
Lab D	74.22	25.93	68.60	21.81	70.14	21.31
	74.09	25.78	68.78	21.59	70.35	21.30

Table 1   Comparison of analytical results.

NO	SP	TP gm%	AB gm%	CA mg%	IP mg%	GL mg%	BN mg%	UA mg%	CT mg%	TB mg%	AP mU/ml	LD mU/ml	GO mU/ml	CK mU/ml	PV %
62	C	7.3	4.8	9.3	4.7	198	14	0.07	0.8	0.2	187	242	40	71	40
	1H	—	—	—	—	—	—	—	—	—	—	—	—	>778	38
	48H	8.0	4.6	10.0	4.1	93	16	0.15	0.9	0.2	227	330	48	183	38
86A	C	7.7	3.9	9.0	2.6	116	11	0.29	1.0	0.1	152	226	18	<27	43
	1H	7.6	3.9	9.2	2.4	97	14	0.31	1.0	0.2	167	480	42	192	43
	24H	8.2	3.5	10.1	5.9	149	16	0.29	1.3	0.3	167	336	36	250	38
71	C	8.3	3.3	9.8	3.0	105	20	0.20	0.9	0.2	580	175	20	40	41
	1H	8.4	3.6	10.1	3.5	108	21	0.15	0.9	0.2	610	439	38	830	42
	24H	8.1	4.0	10.0	4.6	122	12	0.19	0.9	0.2	>350	294	48	380	37
117	C	8.1	4.0	9.3	5.4	81	14	0.30	0.8	0.2	110	660	48	58	40
	1H	7.4	3.8	9.2	5.0	78	15	0.19	0.8	0.1	107	493	47	500	40
	24H	7.6	3.7	9.7	5.7	185	15	0.47	1.1	0.2	135	381	51	220	37
33	C	7.2	3.7	8.8	2.4	119	15	0.28	1.1	0.2	128	317	50	67	39
	1H	7.4	3.8	9.4	4.0	80	15	0.19	0.8	0.2	156	569	66	640	42
	24H	7.8	3.9	10.0	4.2	109	21	0.25	1.1	0.2	170	373	57	245	39
58	C	8.8	5.7	10.5	4.6	77	16	0.19	1.1	0.2	194	583	86	>778	46
	1H	7.5	5.2	10.1	4.3	72	16	0.09	0.8	0.3	195	820	113	>778	43
	24H	7.5	4.5	9.5	4.6	225	27	0.28	1.5	0.6	232	>600	>300	>788	31
74	C	7.2	4.9	9.5	4.7	75	18	0.12	1.1	0.2	138	195	38	39	39
	1H	6.8	4.5	9.0	4.5	36	18	0.10	0.9	0.3	140	252	44	210	35
	48H	7.6	4.2	9.6	6.0	122	19	0.19	1.1	0.1	186	224	40	435	37
113	C	7.3	3.9	8.9	3.8	72	22	0.17	0.8	0.2	223	192	53	52	40
	1H	7.2	3.7	8.9	3.4	70	24	0.12	0.7	0.2	234	109	60	144	42
	24H	7.6	4.1	9.6	2.9	111	16	0.21	0.8	0.3	236	207	83	86	38

Table 2   Blood chemistry before and after liver biopsy and before test dose or infusion with PFDE in the monkey.

continued on Table 3

NO	SP	TP gm%	AB gm%	CA mg%	IP mg%	GL mg%	BN mg%	UA mg%	CT mg%	TB mg%	AP mU/ml	LD mU/ml	GO mU/ml	CK mU/ml	PV %
80*	C	7.4	4.5	9.9	4.4	87	19	0.1	1.0	0.2	122	185	38	<27	45
	24H	7.6	4.0	10.0	12.0	176	24	0.5	1.3	0.2	177	380	78	835	43
95	C	6.9	4.4	9.0	1.3	111	16	0.3	1.0	0.2	65	191	25	<25	42
	1H	6.9	4.3	9.0	2.9	97	17	0.2	0.9	0.2	68	388	51	210	45
	24H	7.2	4.3	9.3	2.1	122	15	0.2	1.1	0.2	83	321	96	300	39
98	C	7.9	4.5	9.0	4.0	92	14	0.1	1.0	0.1	97	410	38	27	42
	1H	7.5	4.4	8.6	5.0	62	15	0.2	0.8	0.2	108	509	55	280	38
	24H	8.0	4.6	9.5	3.3	110	15	0.1	1.1	0.2	99	282	38	220	40
97A	C	8.1	4.9	9.3	2.9	104	18	0.2	1.5	0.2	75	204	33	<25	44
	1H	7.8	4.7	9.1	3.2	66	19	0.2	1.2	0.4	80	570	76	280	46
	24H	8.4	5.1	10.7	3.3	119	17	0.1	1.2	0.1	105	364	51	210	44
Mean	C	7.7	4.4	9.4	3.6	103	16	0.20	1.0	0.2	171	298	41	51	42
	1H	7.4	4.2	9.3	3.8	77	17	0.17	0.9	0.2	186	463	59	365	41
	24H	7.8	4.2	9.8	4.9	143	18	0.27	1.1	0.2	156	326	60	305	39
S.E.	C	0.17	0.19	0.15	0.36	10	0.9	0.02	0.06	0.01	45	50	5.5	6.5	0.7
	1H	0.14	0.19	0.16	0.29	7	1.1	0.02	0.05	0.03	52	64	7.4	84	1.1
	24H	0.14	0.15	0.14	0.93	13	1.6	0.04	0.07	0.04	20	20	7.4	75	1.2

Table 3   Blood chemistry before and after liver biopsy and before test dose or infusion with PFDE in the monkey.

C = sample taken from the anesthetized monkey before biopsy. IH = sample one hour after incision closed. 24 and 48H = 24 and 48 hours after incision closed.

*bone marrow biopsied.

| | MONKEYS | | | | DOGS | | |
MK NO	Before	After 1	After 2	DOG NO	Before	After 1	After 2
62	101/50	107/52	97/50	—	—	—	—
86A	111/61	120/51	103/51	13	145/135	34/22	120/115
71	124/50	129/52	131/55	ALNO	160/135	50/38	—
117	117/49	118/52	126/50	ZSASE2	148/114	65/44	143/120
33	143/74	147/70	151/71	—	—	—	—
58	140/64	143/66	146/70	—	—	—	—
74	108/49	118/49	114/50	ZSASE2	148/114	65/44	143/120
113	124/51	130/51	136/51	BKYO	110/82	22/7	110/86
"	"	"	"	YK4	132/109	23/15	104/93
80	105/85	110/85	120/85	—	—	—	—
95	116/52	116/51	119/52	ZSPSE3	131/124	50/40	140/123
98	135/62	139/61	127/60	ZY	170/127	40/33	165/135
97A	124/49	125/52	119/50	VVEO	125/93	36/25	123/101
—	—	—	—	8316136	105/93	22/13	120/107
—	—	—	—	4915439	130/125	45/38	127/123
Mean	121/58	125/58	124/58		136/114	39/27	128/111
S.E.	4.1/3.5	3.9/3.3	4.8/3.5		6.8/6.3	4.7/4.3	6.6/5.6

Table 4 Comparison of the effect on arterial pressure of two injections of 0.05 ml/kg of emulsion in the monkey and dog. The pressures are systolic/diastolic in mm. of mercury. Dogs opposite monkeys on the table were given the same emulsion on the same day. *Two dogs (BKYO and YK4) were tested the same day as monkey 113.

NO	TP	AB	CA	IP	GL	BN	UA	CT	TB	AP	LD	GO	CK	NA	K	CL	CO_2	CH	PV
98	7.9	4.4	7.9	3.1	84	18	0.16	0.9	0.2	112	395	39	21	138	6.9	103	14	157	42
	7.7	4.2	9.2	3.3	77	18	0.11	0.9	0.2	106	370	38	38	138	5.6	104	18	145	42
113	8.3	4.9	9.2	4.0	90	18	0.19	0.9	0.2	157	148	42	104	151	4.3	108	25	133	43
74	8.2	4.6	9.1	4.4	96	20	0.38	0.9	0.2	135	297	58	450	145	4.9	106	14	162	42
	7.5	4.3	8.7	3.6	111	17	0.29	1.0	0.2	192	280	53	91	148	4.8	104	13	133	42
62	7.5	4.1	7.8	3.3	99	18	0.29	0.9	0.2	156	312	54	86	141	6.3	105	16	129	40
	8.1	4.5	8.1	3.9	82	15	0.20	0.9	0.3	169	404	47	87	143	6.5	107	13	143	45
95	7.9	4.5	8.6	4.0	88	16	0.28	1.0	0.3	150	476	50	188	139	5.4	105	15	134	44
	7.5	4.1	8.6	3.6	69	15	0.12	0.9	0.2	84	351	44	50	141	4.6	104	18	167	41
	7.3	3.8	8.4	3.3	74	14	0.20	0.9	0.2	77	383	49	30	138	4.8	103	20	148	42
71	7.7	4.2	9.0	3.8	62	15	0.15	0.9	0.2	130	310	52	83	144	5.3	107	13	210	45
	7.6	4.1	9.1	3.4	85	15	0.20	1.0	0.2	128	324	60	124	142	4.5	106	18	207	42
58	7.5	4.3	9.4	2.5	58	18	0.29	0.8	0.2	110	347	42	42	148	5.0	108	20	175	43
	7.9	4.3	10.0	3.2	67	14	0.19	1.0	0.2	188	250	47	70	150	3.6	110	20	156	44
33	7.8	4.1	9.1	2.8	74	17	0.30	0.9	0.3	119	306	39	18	148	5.4	109	14	170	42
	8.0	4.3	9.6	3.2	94	14	0.21	1.1	0.2	133	166	30	40	150	3.8	115	12	145	43
97A	8.0	4.4	10.2	3.6	89	19	0.29	0.9	0.2	78	427	81	188	151	5.5	110	10	170	41
	8.3	4.5	9.8	4.1	82	14	0.17	1.0	0.1	80	254	37	40	150	3.7	108	13	144	42
86A	8.0	4.3	9.6	2.0	77	14	0.20	0.7	0.2	119	252	30	15	150	5.4	107	18	190	43
	8.2	4.5	9.8	2.8	85	12	0.15	0.9	0.1	124	155	25	32	152	4.0	110	19	159	43
117	7.6	3.9	7.9	4.3	104	15	0.31	0.6	0.2	73	323	27	16	150	5.1	111	9	460	39
	7.9	4.1	9.6	4.4	90	16	0.20	0.8	0.1	73	240	28	34	152	4.3	113	13	159	42
Mean	7.8	4.3	9.0	3.5	84	16	0.22	0.9	0.2	122	308	44	84	146	5.0	107	16	173	42
S.E.	0.06	0.05	0.15	0.14	2.9	0.45	0.02	0.02	0.01	7.9	19	2.8	21	1.1	0.19	0.70	0.84	14.8	0.07

Table 5 Analysis of blood and plasma from awake monkeys after biopsy and blood pressure test dose but before PFDE infusion

NO	pO₂ AORTIC B	pO₂ AORTIC D	pO₂ AORTIC A	pO₂ VENOUS B	pO₂ VENOUS D	pO₂ VENOUS A	pCO₂ AORTIC B	pCO₂ AORTIC D	pCO₂ AORTIC A	pCO₂ VENOUS B	pCO₂ VENOUS D	pCO₂ VENOUS A	pH AORTIC B	pH AORTIC D	pH AORTIC A	pH VENOUS B	pH VENOUS D	pH VENOUS A
C1	378	455	369	43	68	56	36	39	33	42	39	36	---	---	---	7.30	7.27	7.28
C4	400	415	478	40	53	58	36	38	41	44	47	47	7.34	7.34	7.34	7.33	7.23	7.26
C5	360	484	524	39	62	81	41	39	39	44	47	48	7.36	7.32	7.32	7.37	7.31	7.30
62	311	436	452	46	50	61	35	35	33	45	30	44	7.40	7.34	7.41	7.48	7.42	7.43
86A	378	416	489	33	43	59	30	26	26	37	33	35	7.52	7.48	7.54	7.43	7.37	7.38
71	369	420	318	34	57	52	43	38	39	45	50	45	7.44	7.46	7.43	7.45	7.42	7.37
117	416	461	494	31	45	63	25	28	30	34	40	40	7.52	7.49	7.44	7.35	7.31	7.37
33	227	407	306	45	57	71	29	34	34	35	43	37	7.39	7.38	7.41	7.36	7.46	7.46
58	341	430	448	33	38	52	28	32	34	36	42	42	7.43	7.54	7.54	---	---	---
C2	332	414	440	36	64	135	42	36	30	46	46	41	---	---	---	---	---	---
C3	295	352	354	38	69	87	40	39	36	43	45	42	---	7.41	7.50	7.43	7.36	7.44
74	379	466	494	42	122	233	23	23	35	28	26	42	7.42	7.41	---	7.40	7.40	7.43
113	---	---	---	48	150	233	---	---	---	22	23	22	---	---	---	---	---	---
Mean	349	430	430	39	68	95	34	34	34	39	39	40	7.42	7.42	7.44	7.39	7.36	7.37
S.E.	15.7	10.5	22.5	1.6	9.3	19	2.1	1.7	1.3	2.1	2.5	1.9	0.22	0.28	0.28	0.19	0.25	0.24

Table 6   Blood gases and pH before, during, and after infusion of PFDE in the monkey.

"Aortic" refers to samples removed from the cannula inserted into the aorta via the femoral artery. "Venous" refers to mixed venous samples removed from a catheter either in the right ventricle or the pulmonary artery. Blood gas tensions are in mm. of mercury. B = before, D = during infusion, A = after. The average uncorrected barometric pressure in Cincinnati is 744 mm Hg.

DATE	TP gm%	AB gm%	CA mg%	IP mg%	GL mg%	BN mg%	UA mg%	CT mg%	TB mg%	AP mU/ml	LD mU/ml	GO mU/ml	CK mU/ml	CH mg%
11-08-73	7.4	4.8	9.6	3.6	---	14	7.5	---	0.4	71	205	64	---	235
12-07-73	7.9	5.0	9.5	2.8	81	14	7.8	---	0.4	71	215	88	---	213
01-09-74	8.1	4.8	9.6	3.6	94	18	6.8	---	0.4	57	221	74	---	252
05-08-74	7.6	5.1	9.9	3.6	111	11	8.7	1.0	0.4	52	222	67	---	225
06-27-74	8.1	5.3	10.0	3.4	133	11	8.5	1.0	0.3	55	220	70	---	248
07-24-74	7.8	5.2	10.1	3.5	111	9	9.2	1.1	0.4	61	201	69	---	247
09-03-74	7.8	5.0	9.8	3.3	100	10	7.7	1.1	0.5	58	217	86	---	244
10-10-74	7.1	5.0	9.3	3.3	77	15	8.7	1.0	0.4	64	166	68	---	206
11-09-74	7.7	5.1	10.1	4.1	97	11	7.7	1.1	0.5	56	210	68	---	254
12-09-74	7.6	5.5	9.6	3.6	130	10	7.1	1.1	0.5	72	219	82	82	---
03-19-75	7.3	4.4	9.5	3.1	119	12	6.9	0.9	0.5	53	225	92	104	220
07-10-75	8.0	4.9	9.9	3.3	125	12	8.1	1.0	0.6	57	260	138	40	250
08-14-75	7.6	4.6	9.6	2.8	118	11	8.1	1.0	0.5	60	190	72	36	221
Mean	7.60	4.97	9.80	3.4	108	12.2	7.9	1.0	0.4	60.5	206	79.8	65.5	235
S.E.	0.30	0.28	0.22	0.35	18.3	2.5	0.73	0.06	0.07	06.9	3.6	01.9	33.0	16.8
Normal Range	6.0 8.0	3.5 5.0	8.5 10.5	2.5 4.5	50 120	10 20	2.0 7.0	0.7 1.5	0.2 1.0	25 125	100 225	7 40	15 110	150 300

Table 7   Random blood samples from L.C. used as laboratory control for the monkeys.

| MK NO | HEART RATE (BEATS/MINUTE) | | | RESPIRATION RATE (BREATHS/MINUTE) | | |
	BEFORE INFUSION	DURING INFUSION	AFTER INFUSION	BEFORE INFUSION	DURING INFUSION	AFTER INFUSION
C1	207	200	210	42	66	87
C4	216	203	212	37	42	52
C5	213	200	198	32	47	42
62	193	148	140	36	36	51
86A	187	189	170	25	31	36
71	209	190	171	25	58	84
117	238	205	179	30	40	48
33	212	189	186	41	43	42
58	213	186	179	31	36	42
Mean	210	190	183	33	44	54
S.E.	5.10	6.09	7.90	2.20	3.96	6.60
C2	214	201	180	34	39	51
C3	174	156	152	70	77	72
74	170	174	162	23	32	34
113	206	176	164	47	65	70
Mean	191	177	165	44	53	57
S.E.	12.8	10.7	6.7	11.7	12.3	10.3
Mean	204	186	177	36	47	55
T-Test	1.85	1.25	1.52	-1.45	-1.01	-0.27

Table 8   Pulse and respiration before during and after infusion of 10% PFDE (C1-58) and 20% PFDE (C2-113)

| MK NO | ARTERIAL PRESSURE mm H g | | | RV OR PA PRESSURE mm H g | | |
	BEFORE INFUSION	DURING INFUSION	AFTER INFUSION	BEFORE INFUSION	DURING INFUSION	AFTER INFUSION
C1	134	126	107	10	14	16
C4	111	122	121	8	16	12
C5	115	120	121	2	9	18
62	88	79	75	3	15	20
86A	111	111	110	8	17	12
71	112	105	101	11	19	19
117	100	108	100	7	19	20
33	88	95	102	12	19	18
58	102	102	96	6	11	9
Mean	107	108	104			
S.E.	5.07	5.19	4.93			
C2	101	111	120	8	14	19
C3	89	93	97	16	18	18
74	88	71	82	8	18	21
113	107	124	124	5	19	19
Mean	96	100	106			
S.E.	5.36	13.28	11.44			
Mean	104	105	104			
T-Test	1.33	0.75	−0.22			

Table 9 Blood pressures in the rhesus monkey before during and after infusion of 10% PFDE (C1-58) and 20% PFDE (C2-13)

NO	SP	TP	AB	CA	IP	GL	BN	UA	CT	TB	AP	LD	GO	CK	CH	PV
62	C	7.1	4.0	8.7	1.8	67	19	0.35	0.8	0.1	168	364	98	65	138	41.0
	1H	—	—	—	—	—	—	—	—	—	—	—	—	—	—	44.0
	24H	8.1	4.7	9.1	2.5	77	15	0.15	1.0	0.1	236	333	105	95	133	43.5
86A	C	7.0	3.7	9.1	3.8	94	14	0.32	0.9	0.1	158	162	32	46	178	40.0
	1H	6.7	3.4	8.7	2.6	91	14	0.21	0.8	0.1	164	212	34	250	175	43.0
	24H	QNS	—	—	—	—	—	—	—	—	—	—	—	90	172	38.5
71	C	7.1	3.8	9.2	3.5	126	22	0.18	1.1	0.1	153	292	38	110	189	39.0
	1H	7.0	3.8	9.1	2.9	89	21	0.20	1.0	0.2	154	420	52	420	196	38.5
	24H	7.8	3.9	9.8	3.4	105	12	0.20	1.1	0.2	152	394	117	550	191	40.0
117	C	7.1	4.0	8.9	4.7	75	15	0.10	0.7	0.1	75	225	23	170	163	42.0
	1H	QNS	—	—	—	—	—	—	—	—	—	—	—	—	—	39.0
	24H	7.2	3.7	9.0	5.3	132	14	0.21	1.0	0.2	156	444	82	196	126	37.0
33	C	6.6	3.6	8.7	3.8	132	16	0.21	1.0	0.2	86	162	30	39	149	39.5
	1H	6.7	3.6	8.8	1.4	68	15	0.15	0.8	0.2	108	249	34	112	159	41.0
	24H	7.8	4.4	10.1	2.6	103	17	0.28	1.2	0.2	102	256	40	82	146	40.0
58	C	QNS	—	—	—	—	—	—	—	—	—	—	—	110	—	34.0
	24H	7.1	4.0	9.8	2.9	96	14	0.28	1.3	0.4	163	302	47	132	165	39.0
74	C	6.7	3.5	8.0	3.6	74	16	0.12	0.6	0.2	207	125	26	54	107	39.5
	1H	5.8	2.9	7.5	2.2	71	16	0.08	0.5	0.2	193	138	26	104	104	44.0
	24H	7.0	4.1	8.5	5.5	128	16	0.22	1.0	0.2	197	352	36	130	102	40.0
113	C	7.1	4.3	8.4	3.9	81	19	0.12	0.7	0.2	116	184	38	56	142	38.0
	1H	6.9	4.2	8.4	4.2	71	19	0.09	0.6	0.2	128	143	38	102	138	42.0
	24H	8.2	4.5	9.1	5.6	125	14	0.30	1.2	0.1	145	174	34	130	141	43.0
80	C	6.3	3.2	8.9	4.0	168	27	0.20	0.8	0.3	122	367	63	>854	84	34.0
	1H	5.0	2.6	14.9	6.6	102	24	0.23	0.7	0.2	104	362	67	854	72	28.0
	24H	6.8	3.4	7.6	10.8	218	70	0.47	2.4	0.4	137	3950	480	847	111	32.0
95	C	7.1	3.3	8.4	3.4	92	16	0.25	0.9	0.1	76	207	40	48	116	41.0
	1H	6.6	3.0	8.4	3.0	78	17	0.13	0.8	0.2	79	286	44	720	133	49.5
	24H	7.0	3.9	9.6	2.3	97	16	0.14	1.2	0.1	88	256	53	112	128	40.0

Table 10 Blood chemistry before and after test dose of PFDE continued on Table 11

NO	SP	TP	AB	CA	IP	GL	BN	UA	CT	TB	AP	LD	GO	CK	CH	PV
98	C	7.0	4.0	8.4	3.0	82	17	0.18	0.8	0.1	86	187	23	30	131	39.0
	1H	6.7	3.8	8.4	1.9	76	17	0.06	0.7	0.2	98	204	28	60	141	44.0
	24H	7.7	4.3	9.0	3.2	89	13	0.09	1.0	0.2	101	275	30	44	119	37.0
97A	C	7.1	3.9	9.3	1.3	76	18	0.21	0.8	0.2	72	189	24	72	143	41.5
	1H	7.1	4.0	8.9	2.4	56	17	0.12	0.9	0.2	64	156	18	38	151	42.0
Mean	C	6.9	3.8	8.7	3.3	97	18	0.20	0.8	0.2	120	224	40	140	140	39.5
	1H	6.5	3.5	9.2	3.0	78	18	0.14	0.8	0.2	121	241	38	277	141	40.8
	24H	7.5	4.1	9.2	4.5	117	15	0.23	1.2	0.2	148	309	60	219	139	39.1
S.E.	C	0.09	0.10	.13	0.31	10	1.2	0.02	0.04	0.02	14	26	7.1	76	9.7	0.69
	1H	0.24	0.19	.77	0.55	5	1.1	0.02	0.05	0.01	15	35	5.2	98	13.0	1.66
	24H	0.17	1.13	.24	0.84	13	0.6	0.04	0.14	0.04	15	29	11.5	79	8.6	0.99

Table 11   Blood chemistry before and after test dose of PFDE

NO	TP	AB	CA	IP	GL	BN	UA	CT	TB	AP	LD	GO	Na	K	Cl	CO_2	HCT
N	28	28	28	28	28	28	27	28	28	28	28	28	26	26	26	26	28
M	8.1	4.5	10.2	5.0	112	19.3	.29	1.17	0.2	390	387	58	160	5.1	112	15	37
SE	.29	.15	.27	.37	4.6	1.21	.02	0.37	.02	25.8	52	11	1.60	0.33	0.91	1.7	1.6

Table 15   Blood chemistry for one year's sampling after infusion of PFDE in Melvin.

NO	SP	TP	AB	CA	IP	GL	BN	UA	CT	TB	AP	LD	GO	CK	CH	PV	PF
C1	1DPP	7.8	3.8	9.1	5.7	140	11	0.50	0.8	0.2	~500	340	30	194	---	42	0
	IPP	7.2	4.0	8.7	7.4	83	12	0.19	0.6	0.2	~500	340	30	148	131	39	0
	PI	4.9	2.6	7.3	7.3	55	10	0.21	0.7	0.2	348	292	24	142	---	28	12
C4	8DPP	8.7	4.1	9.7	5.2	134	12	0.18	0.8	0.1	202	406	33	---	---	44	0
	1DPP	9.2	4.7	11.3	4.5	92	21	0.36	0.9	0.2	182	377	52	66	149	38	0
	IPP	6.0	3.1	8.9	4.4	63	20	0.10	0.6	0.1	158	194	28	88	92	38	12
	PI	5.8	3.0	8.2	6.2	68	18	0.10	0.6	0.2	166	519	51	190	12	20	0
C5	8DPP	7.9	4.2	9.9	4.8	112	18	0.28	0.8	0.2	208	258	31	112	200	44	0
	1DPP	7.8	4.3	10.3	5.5	76	18	0.33	0.9	0.2	200	252	31	114	183	43	0
	IPP	6.8	3.9	9.2	4.6	99	19	0.19	0.6	0.3	207	299	36	210	181	36	0
	PI	2.2	1.2	6.6	4.2	81	14	0.09	0.5	0.1	82	155	17	80	44	14	12
62	1DPP	7.4	4.4	10.0	1.8	72	19	0.20	1.0	0.2	145	313	42	50	156	44	0
	IPP	6.5	3.6	8.4	2.6	85	16	0.20	0.8	0.2	162	286	43	102	122	38	0
	PI	1.7	0.9	5.9	2.9	64	14	0.12	0.7	0.1	47	92	17	38	24	11	13
86A	2HPI	2.1	1.1	6.0	2.5	123	15	0.18	0.9	0.1	69	148	26	80	48	13	13
	1DPP	7.8	4.5	9.4	3.4	82	15	0.28	1.0	0.1	135	162	38	68	220	46	0
	IPP	6.5	3.9	8.4	4.0	97	11	0.09	0.7	0.2	129	183	27	52	152	39	0
	MP	6.0	4.4	8.5	3.6	95	11	0.02	0.3	0.3	69	133	17	38	98	25	0
	PI	3.1	2.2	6.8	3.5	65	10	0.02	0.7	0.3	43	102	10	38	36	13	12
71	1DPI	6.2	3.8	9.2	5.8	125	18	0.34	1.1	0.1	88	1520	325	857	46	13	6
	8DPP	7.2	4.3	9.5	3.0	93	17	0.29	1.0	0.1	104	190	43	26	220	45	0
	IPP	6.0	3.6	8.5	3.2	102	18	0.12	0.6	0.2	114	233	32	62	190	40	0
	MP	6.0	4.6	8.7	3.0	92	16	0.10	0.7	0.3	58	147	17	26	124	24	0
	PI	3.1	2.4	7.2	2.8	69	14	0.08	0.7	0.3	44	117	13	90	55	9	13
117	3DPI	6.1	3.4	8.7	4.8	97	24	0.30	0.9	0.2	230	~1160	183	1502	190	15	4
	1DPP	7.4	4.0	10.0	5.0	102	18	0.01	0.8	0.1	79	170	16	8	182	40	0
	IPP	6.0	3.2	8.2	3.8	93	18	0.04	0.5	0.2	83	200	23	54	145	32	0
	MP	5.9	4.1	8.5	3.7	79	18	0.06	0.6	0.2	52	125	12	36	82	20	0
	PI	3.3	2.2	7.0	3.4	70	15	0.02	0.6	0.2	68	94	8	40	34	8	13
	2DPI	6.3	3.4	9.1	3.8	122	16	0.21	0.8	0.2	214	>600	265	14660*	123	17	4

Table 12   Blood chemistry before and after 10% PFDE. Continued on Table 13

NO	SP	TP	AB	CA	IP	GL	BN	UA	CT	TB	AP	LD	GO	CK	CH	PV	PF
33	7DPP	8.0	4.5	9.7	3.2	83	16	0.11	0.9	0.1	116	137	28	8	166	44	0
	IPP	5.9	3.1	8.2	4.0	76	16	0.10	0.9	0.1	137	273	28	232	130	36	0
	MP	5.7	3.5	8.4	3.9	66	15	0.11	0.9	0.1	73	174	18	134	78	22	0
	PI	3.3	2.2	7.1	3.5	42	14	0.15	0.9	0.2	54	213	15	144	34	12	11
	3DPI	—	—	—	—	—	—	—	—	—	—	—	—	—	—	11	1
58	2DPP	6.7	2.4	7.7	4.0	53	14	0.23	0.9	0.1	258	256	105	90	220	48	0
	IPP	5.7	3.1	8.9	1.5	82	17	0.19	0.9	0.3	199	408	94	374	153	35	0
	MP	5.5	3.9	9.2	1.2	82	15	0.21	0.9	0.5	122	258	57	240	87	23	0
	PI	2.9	2.2	7.2	0.9	58	13	0.15	0.8	0.4	62	207	42	204	23	9	10
	2DPI	5.9	3.3	10.0	4.4	89	19	0.40	1.1	0.4	237	1740	412	3640	70	8	4
Mean	DPP	7.8	4.1	9.7	4.2	94	16	0.25	0.89	0.15	194	260	41	74	188	44	0
	IPP	6.3	3.5	8.6	3.9	87	16	0.14	0.69	0.20	188	268	38	147	144	37	0
	MP	5.8	4.1	8.7	3.1	83	15	0.10	0.68	0.28	75	167	24	95	94	23	0
	PI	3.2	2.9	6.9	3.8	70	14	0.11	0.71	0.21	98	194	22	105	34	14	12.1
	DPI	6.1	4.1	9.2	4.7	108	19	0.28	0.98	0.23	192	1473	296	6601	107	13	3.8
S.E.	DPP	0.22	0.20	0.28	0.39	8.3	1.0	0.04	0.03	0.02	36	29	7	19	10	0.9	0
	IPP	0.17	0.13	0.12	0.57	4.4	1.1	0.02	0.05	0.02	44	26	8	38	11	0.9	0
	MP	0.11	0.22	0.16	0.55	5.8	1.3	0.04	0.12	0.07	14	27	9	46	9	1.0	0
	PI	0.42	0.75	0.22	0.11	7.1	0.8	0.02	0.04	0.03	32	44	5	21	5	2.0	0.3
	DPI	0.10	0.82	0.31	0.49	10.4	2.0	0.05	0.09	0.07	41	207	56	4993	37	1.8	0.9

Table 13   Blood chemistry before and after infusion of 10% PFDE in 9 monkeys.

(DPP = days pre-phlebotomy, IPP = immediately pre-phlebotomy, MP = mid-phlebotomy, PI = immediately post-infusion, HPI = hours post-infusion, DPI = day (post-infusion), *This value is considered erroneous and has been eliminated from the mean and standard error.

NO	SP	TP	AB	CA	IP	GL	BN	UA	CT	TB	AP	LD	GO	CK	CH	PV	PF
C2	IDPP	7.4	3.8	8.6	5.8	145	11	0.31	0.7	0.1	207	327	33	690	117	39	0
	IPP	5.1	2.7	8.1	5.0	88	13	0.09	0.6	0.2	174	258	32	410	78	32	0
	PI	1.9	1.0	6.3	5.6	79	13	0.09	0.6	0.1	68	124	17	178	22	32	11
C3	IDPP	8.5	4.5	10.3	5.0	122	14	0.21	0.8	0.2	226	302	29	96	195	45	0
	IPP	4.6	2.5	8.0	5.2	83	15	0.09	0.6	0.2	175	180	21	92	106	24	0
	PI	1.6	0.9	6.3	5.6	65	15	0.09	0.7	0.2	67	95	12	46	44	15	54
74	IDPP	7.7	3.8	9.6	2.1	80	19	0.2	1.1	0.2	138	165	39	18	134	44	0
	IPP	6.3	3.5	8.4	3.1	80	19	0.2	0.8	0.3	143	171	34	54	124	31	0
	MP	5.9	4.1	8.5	3.0	72	18	0.1	0.7	0.3	88	118	22	36	97	20	0
	PI	2.6	1.9	6.6	3.3	38	19	0.1	0.9	0.4	51	75	13	44	35	15	32
113	IDPP	7.9	4.4	10.4	4.0	104	21	0.3	1.0	0.3	124	276	44	236	189	44	0
	IPP	6.8	3.7	8.6	4.5	75	21	0.2	0.9	0.5	137	279	51	348	157	37	0
	MP	6.5	2.3	9.8	4.8	31	12	0.7	0.7	0.4	65 >	600	65	240	100	23	0
	PI	6.5	4.5	8.8	4.4	69	19	0.2	0.9	0.7	81	169	30	250	97	11	23
	IDPI	6.3	4.4	6.2	>10.	117	61	5.6	6.1	0.8	181	> 600	315	>24000	46	16	24
Mean	DPP	7.9	4.1	9.7	4.2	113	16	0.26	0.9	0.2	174	268	36	260	159	43	0
	IPP	5.7	3.1	8.3	4.5	82	17	0.15	0.7	0.3	157	222	34	226	116	31	0
	MP	6.2	3.2	9.2	3.9	52	15	0.40	0.7	0.4	76	-	44	138	98	22	0
	PI	3.2	2.1	7.0	4.7	63	16	0.12	0.8	0.4	67	116	18	130	50	18	30
S.E.	DPP	0.27	0.22	0.48	0.92	16	2.6	0.03	0.1	0.1	29	41	4	174	23	1.6	0
	IPP	0.59	0.34	0.16	0.55	3	2.1	0.04	0.1	0.1	12	32	7	103	19	3.1	0
	MP	0.42	1.27	0.92	1.27	29	4.2	0.42	0.0	0.1	16	-	30	144	2	2.1	0
	PI	1.31	0.34	0.70	0.64	10	1.7	0.03	0.1	0.2	7	24	5	59	19	5.4	10

Table 14 Blood chemistry before and after infusion of a 20% PFDE. Any values designated by > were excluded from the statistics.

(DPP = days pre-phlebotomy, IPP = immediately pre-phlebotomy, MP = mid-phlebotomy, PI = immediately post-infusion, HPI = hours post-infusion, DPI = day post-infusion)

NO	WT	PHLEBOTOMY				INFUSION				4 HR. POST INFUSION				OVER 4 HR. POST INFUSION			TOTALS IN AND OUT					
		OUT BL	IN RI	IN AB	TO	OUT BL	IN PFDE	IN RI	TO	OUT BL	IN RI	IN OS	TO	IN RI	IN OS	TO	OUT BL	IN RI	IN OS	PFDE	TO	BAL
C1	3.2	10	4	0	4	6	60	9	69	18	4	0	4	0	0	0	34	17	0	60	77	43
C4	3.2	28	54	0	54	4	50	5	55	3	5	0	5	0	0	0	35	64	0	50	114	79
C5	4.4	26	55	0	55	3	50	3	53	6	3	0	3	0	0	0	35	61	0	50	111	76
62	6.6	26	53	0	53	2	50	2	52	7	12	19D	31	0	11D	11	35	68	30D	50	148	113
86A	6.1	30	3	29	32	2	50	3	53	4	2	3A	5	0	0	0	36	8	31A	50	89	53
71	8.0	26	4	25	29	2	50	2	52	3	3	0	3	0	0	0	31	9	25A	50	84	53
117	6.8	16	3	21	24	2	50	3	53	4	1	3A	4	7	9A	16	32	14	33A	50	97	65
33	7.4	26	2	33	35	2	50	2	52	3	1	0	1	0	0	0	21	5	33A	50	88	67
58	7.1	26	5	25	30	2	50	3	53	3	1	0	1	0	0	0	31	9	25A	50	84	53
C2	3.0	26	49	0	49	4	50	6	56	1	5	0	5	0	23D	23	41	60	23D	50	133	92
C3	3.8	26	53	0	53	3	50	3	53	8	9	16D	25	0	0	0	37	65	16D	50	131	94
74	6.2	25	5	30	35	2	50	3	53	8	27	16A	43	16	8A	24	35	51	54A	50	155	120
113	5.6	26	7	25	32	1	50	3	53	4	0	0	0	0	0	0	31	10	25A	50	85	54

Table 16 Blood out and fluids in before, during and after infusion of PFDE.

Monkey weight is in kilograms. Blood out and fluids in are expressed as ml/kg. OS = osmotic or oncotic, D = Dextran 40, A = albumin, BL = blood, RI = Ringer's solution, TO = total, Bal = balance or total in minus total out.

NO.	EMULSION RECEIVED	BODY WEIGHT	LIVER	SPLEEN	LUNG	KIDNEYS
	%PFDE	kg				
62	10	7.2	26.3	1.1	11.8	3.7
117	10	6.4	22.9	1.9	7.2	4.4
C2	20	3.0	47.5	2.3	12.3	5.7
C3	20	4.2	37.7	6.3	13.6	5.1
74	20	6.4	32.3	1.2	14.9	5.0
113	20	5.7	29.0	1.5	9.9	4.6
M	-	5.5	32.6	2.4	11.6	4.8
S.E.	-	0.7	4.0	0.9	1.2	0.3
6	0	7.2	17.0	0.7	5.1	3.4
80	0	4.0	36.6	0.9	8.8	-
C7	0	3.9	26.8	1.0	13.3	5.8
M	-	5.0	26.8	0.9	9.1	4.6
S.E.	-	1.3	6.9	0.1	2.9	1.7
FAS	0	3.6	19.2	-	19.0	-

Table 17   Organ weights obtained at autopsy. FAS = normal organ
weights for the rhesus monkey reported by FASEB (20).
All organ weights expressed as gms/kg of body weight.

NO.	DAYS POST INFUSION	PFD INFUSED	PFD FOUND		GLC
		gm	gm	%	
62	1	64	7.7	12	8,000
74	1	120	24	20	12,000
113	1	109	14	13	1,400
C3	8	74	17	23	7,000
C2	14	58	13	22	10,000
C1	42	37	2.1	5.6	1,500
C4	47	31	1.3	4.2	1,700
C5	56	43	2.1	4.8	1,800
86A	64	59	0.3	0.5	4
71	70	78	3.4	4.4	1,000
117	77	66	0.96	1.0	800
58	84	69	0.2	0.3	40
ME	332	25	0.1	0.4	1

Table 18   PFD content of liver obtained at biopsy or autopsy as
determined by sodium biphenyl combustion or vapor phase
GLC.

Data on Monkeys C1, C4, C5, 86A, 71, 58 and ME were
obtained from samples obtained at biopsy; the liver was
assumed to be 3% of the body weight. Relative peak
area was obtained from a pen writing integrator on the
strip chart.

Literature cited

1. Clark, Jr., Leland C. and Gollan, Frank. Science (1966) 152, 1755-1756.
2. Clark, Jr., Leland C., Editor. Federation Proceedings (1970) 29, 1696-1820.
3. Sloviter, Henry A. Medical Clinics of North America (1970) 54, 787-795.
4. Geyer, Robert P. The New England Journal of Medicine (1973) 289, 1077-1082.
5. Clark, Jr., Leland C.; Kaplan, Samuel; Emory, Carolyn and Wesseler, Eugene P. "Progress in Clinical and Biological Research, Vol. I. Erythrocyte Structure and Function", edited by George J. Brewer, 589-600, Alan R. Liss, Inc., New York (1975).
6. Gollan, Frank and Clark, Jr., Leland C. The Alabama Journal of Medical Sciences (1967) 4, 336-337.
7. Gollan, Frank and Clark, Jr., Leland C. The Physiologist (1966) 9, 191.
8. Gollan, Frank and Clark, Jr., Leland C. Transaction of the Association of American Physicians (1967) 80, 102-110.
9. Clark, Jr., Leland C.; Kaplan, Samuel; Becattini, Fernando and Benzing III, George. Federation Proceedings (1970) 29, 1764-1770.
10. Spitzer, Hugh L.; Sachs, George and Clark, Jr., Leland C. Federation Proceedings (1970) 29, 1746-1750.
11. Clark, Jr., Leland C.; Kaplan, Samuel and Becattini, Fernando. Presented at the American Association for Thoracic Surgery 50th Annual Meeting, Washington, D.C. (Abstract No. 34), April 8, 1970.
12. Clark, Jr., Leland C.; Kaplan, Samuel and Becattini, Fernando. Pediatric Research (1970) 4, 464 (Abstract 113).
13. Clark, Jr., Leland C.; Kaplan, Samuel and Becattini, Fernando. The Journal of Thoracic and Cardiovascular Surgery (1970) 60, 757-773.
14. Clark, Jr., Leland C.; Becattini, Fernando and Kaplan, Samuel. Triangle (1972) 11, 115-122.
15. Clark, Jr., Leland C.; Becattini, Fernando and Kaplan, Samuel. The Alabama Journal of Medical Sciences (1972) 9, 16-29.
16. Clark, Jr., Leland C.; Becattini, Fernando; Kaplan, Samuel; Obrock, Virginia; Cohen, David and Becker, Charles. Science (1973) 181, 680-682.
17. Clark, Jr., Leland C.; Wesseler, Eugene P.; Kaplan, Samuel; Miller, Marian L.; Becker, Charles; Emory, Carolyn; Stanley, Lilam; Becattini, Fernando and Obrock, Virginia. Federation Proceedings (1975) 34, 1468-1477.
18. Miller, Marian L.; Clark, Jr., Leland C.; Wesseler, Eugene P.; Stanley, Lilam; Emory, Carolyn and Kaplan, Samuel. The Alabama Journal of Medical Sciences (1975) 12, 84-113.
19. Wesseler, Eugene P.; Iltis, Ron and Clark, Jr., Leland C. The solubility of oxygen in highly perfluorinated liquids. In preparation.

20. Altman, Philip L. and Dittmer, Dorothy S., Editors.
"Biological Handbooks: Growth including reproduction and morphological development". Federation of American Societies for Experimental Biology, Washington, D.C. (1962).
21. Technicon SMA12 procedures were used for most of these analyses. The exact methods used are on file here.
22. Calderwood, H. W.; Modell, J. H.; Rogow, L.; Tham, M. K. and
Hood, C. I. Anesthesiology (1973) 39, 488-495.
23. Wretlind, Arvid. Journal of Nutrition, Metabolic Diseases and
Dietetics (1972) 14, 1-57.
24. Wretlind, Arvid. The pharmacological basis for the use of fat
emulsions in intravenous nutrition. Department of Nutrition and
Food Hygiene. The National of Institute of Public Health Stock-
holm 60, Sweden.
25. Galen, R. S. and Gambino, S. R. Clinical Chemistry (1975)
21, 272.
26. Pooley, F. D.; Kanellopoulos, A. G. and Owen, M. J. Micron
(1972) 3, 486-496.
27. Stein, R. J.; Richter, W. R.; Rdzok, E. J.; Moize, S. M. and
Brynjolfsson, G. "Use of nonhuman primates in drug evaluation",
edited by Harold Vogtborg, 187-199, University of Texas Press,
Austin (1968).
28. Here we have given a summary of the procedure used for making
emulsion. A precise detailed account for making a variety of PFC
emulsions is on file and can be provided on request.
29. Busey, William and Willner, Howard. Normal biochemical parameters of rhesus monkeys and beagle dogs. Presented at the
Technicon Symposium, "Automation in Analytical Chemistry," New
York, N.Y., October 4, 1967. Published by Technicon Corporation.
30. Technicon Industrial System brochure entitled, "The Technicon
SMA 6/60 micro multichannel biochemical analyzer," August 23, 1974.
Technicon number 4229-9-4-2.

Discussion of the paper of Dr. Leland C. Clark, Jr.

Q. How long does it take for the pulmonary arterial pressure to return to normal?

A. About an hour.

Q. So that it was back to normal before the material was cleared from the blood?

A. Yes, it takes over a day before it is cleared from the blood.

Q. Do you have any studies on the perfluoromethyladamatane structure you showed?

A. We have made preliminary studies on a small batch which was somewhat impure but we found that it forms a stable emulsion and also leaves the liver.

Q. In the abstract you mentioned brominated and iodinated perfluorocarbons as X-ray contrast agents. Would you like to comment on that?

A. Dr. Long will soon tell us the story of the X-ray contrast agents. We have found that the iodinated compounds are very little, if any, more X-ray opaque than the brominated compounds, at 55KV. The iodinated compounds are unstable in the presence of light and even though the liberated iodine could be removed by metallic silver there would be other decomposition products there. We have just about given up on iodine.

Q. How do the iodinated compounds hold up in the animal?

A. If they decompose in the presence of light you could almost count on their breakdown in the body. We've had some animals survive small doses. Some of this information has been published. (Clark, Leland C., Jr., Eugene P. Wesseler, Samuel Kaplan, Marian L. Miller, Charles Becker, Carolyn Emory, Lilam Stanley, Fernando Becattini and Virginia Obrock. Emulsions of perfluorinated solvents for intravascular gas transport. Federation Proceedings 34, 1468-1477 (1975); Clark, Leland C., Jr., Eugene P. Wesseler, Marian L. Miller and Samuel Kaplan. Ring versus straight chain perfluorocarbon emulsions as artificial blood. Journal of Microvascular Research 8, 320-340 (1974)).

Q. What are your limitations on volatility of the perfluorocarbons?

A. When used as artificial blood the vapor pressure must be below 50 torr at 38°C. Higher vapor pressures, perhaps <u>much</u> higher vapor pressures can be well tolerated for liquid breathing if 100% oxygen is used as the gas phase.

Q. Do you think these compounds will ever be useful in man?

A. If I didn't I wouldn't be working so hard. This has been a major effort of ours for over eight years.

Q. How about toxicity?

A. When perfluorodecalin is purified by careful distillation in a spinning band column the toxicity decreases to the point where the LD50 of a 10% by volume emulsion is over 200 ml/kg. 200 ml/kg is about three times the blood volume of the mouse. The lower boiling fractions have a lower LD50 than the material as received.

Assessment of the toxicity of PFC emulsions is complicated by the fact that it is still difficult to characterize the final product with a high degree of certainty. Hence one is often unsure whether differences are due to the physical properties of the emulsions themselves or to the various impurities in different PFC.

Generation of fluoride ion by sonication, a former complication of emulsion toxicity testing, has been practically eliminated by the discovery by Geyer that by saturating the liquids being sonicated with carbon dioxide, fluoride ion is not generated. Nitrogen, helium, or hydrogen do not work.

Radiopaque Applications of Brominated Fluorocarbon Compounds in Experimental Animals and Human Subjects

D. M. LONG
Department of Radiology, School of Medicine, University of California, San Diego, Calif.
M. S. LIU, G. D. DOBBEN, and P. S. SZANTO
Hektoen Institute for Medical Research, Cook County Hospital, Chicago, Ill. 60612
A. S. ARAMBULO
College of Pharmacy, University of Illinois at the Medical Center, Chicago, Ill. 60612

Interest in the biological application of perfluorocarbon compounds was begun with the imaginative fluid ventilation studies of Clark and Gollan in 1965.[1] In the first decade since the publication of those experiments, there has been a slowly growing expansion of background information on the effects of perfluorocarbons on the biological systems.[2,3] The industrial use of perfluorocarbon compounds has helped by making more molecules available in pure states at reasonable costs. Imaginative biomedical workers have asked where these new compounds could help solve existing medical problems. Our aggressive chemical industry extended itself in seeking new markets in medicine for the products of their laboratories and plants.

Fluid ventilation as a tool for delivering oxygen to damaged lungs was our initial goal in experiments with perfluorocarbon compounds. This application and the dreams of Jacques Cousteau of a fluid breathing diver in the depths of the ocean seem frustrated for the time being by the unsolved problems of the work of fluid ventilation and by the mechanical injury to the lungs resulting from long periods of fluid ventilation.

We became interested in the possibility of developing a radiopaque fluorocarbon molecule that would be less irritating and safer than currently available radiopaque media which can be quite irritating to the lungs.[4] The inert synthetic fluids, either silicone oils or perfluorocarbons, that were available in 1966 did not possess radiopaque properties. Brominated fluorocarbon liquid was found to be radiodense. Iodinated fluorocarbon compounds were found to be chemically unstable and, therefore, unsuitable for biomedical application. Brominated organic compounds were used in the past and were found to be inferior to iodinated compounds when studied with the usual high kilovoltage used in x-rays of humans. It was argued that brominated compounds, although satisfactory for small animal studies, would be unsatisfactory in the larger human subjects. As can be seen in Figure 1, brominated perfluorocarbon is most satisfactory for use in humans. Lower kilovoltage or energies are used to get maximum absorption from

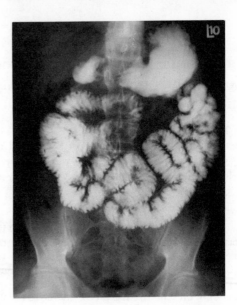

Figure 1. X-ray of the abdomen in a healthy subject ten min after oral administartion of 500 ml of neat radiopaque fluorocarbon. Note the rapid visualization of the jejunal portion of the small intestine.

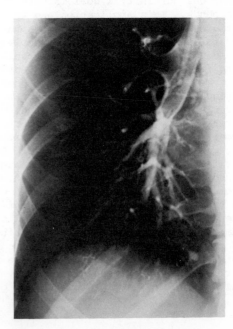

Figure 2. Bronchogram after administration of 6:1 emulsion of radiopaque fluorocarbon in a patient with hemoptysis. A calcified nodule or broncholith can be seen in continuity with an eighth generation bronchus in the lower right hand corner.

the bromine, but this alteration in technique has presented no problems with currently available x-ray equipment. Emulsions of radiopaque fluorocarbon (RFC) were prepared to obtain a material with a higher viscosity to produce a thicker coating of the tracheobronchial tree (Figure 2). The emulsions were prepared in high concentrations of fluorocarbon in physiologic salt solution with Pluronic F-68 as the emulsifying agent and were also non-irritating.

Although we have examined a number of brominated fluorocarbon molecules,[4] most of our studies have been performed with perfluoroctylbromide. This compound is biologically inert and possesses a very low toxicity. The LD_{50} (Table I) of $C_8F_{17}B_R$ is greater than 64 ml./kg. when administered into the gastrointestinal tract. We have used higher dosages of as much as 128 ml./kg. without adverse effects, but these experiments are facetious since the gastrointestinal tract was loaded and overflowing at one or both ends before all the dose was administered. The LD_{50} of neat liquid $C_8F_{17}B_R$ when injected into the lungs was also very high. In fact, animals have been completely submerged in $C_8F_{17}B_R$ and breathed this fluid for short periods of time with survival. The LD_{50} of the 10:1 emulsion of $C_8F_{17}B_R$ in the lungs was greater than 4 ml./kg. Repetitive dosage programs have been performed in animal experiments with no adverse effects. The efficacious doses of the RFC in human diagnostic x-ray studies are given in Table I. There is obviously a wide margin of therapeutic safety when RFC is used in these areas of application.

Gastroenterography

Our initial toxicological and diagnostic studies were directed toward examination of the effects of RFC in the GI tract. It should be remembered that perfluorocarbon compounds are new in biomedical fields, and none of this family of compounds had ever been administered purposefully in large doses to humans. The laboratory studies indicated RFC would be among the safest of drugs or diagnostic agents. When given to experimental animals and human subjects by the GI tract, there have been no adverse effects. Specifically, there was no change in the blood counts, serum enzymes, and urinalysis. The animals showed no change in growth patterns even when given repetitive high doses over short periods of time or over prolonged intervals. RFC has been given repetitively in newborn animals until they achieve young adulthood and to adult animals.

The RFC is odorless and tasteless, and has a low viscosity so that it is easy to drink. Because of the excellent wettability, the mouth and oropharynx are coated immediately after taking the material. Some subjects have complained of an unpleasant oily feeling in the mouth, but most have had no such response. RFC traverses the GI tract more rapidly than food or other contrast agents. When RFC is mixed with food and fed to dogs, the RFC

Table I

	LD$_{50}$ (Animals)	Efficacious Dose
Gastroenterology	> 64 ml./kg. Dogs Rats	1-6 ml./kg.
Alveolography	> 35 ml./kg. Hamsters > 12 ml./kg. Cats	1-2 ml./kg.
Bronchography	> 4 ml./kg. Dogs	0.3-0.6 ml./kg.

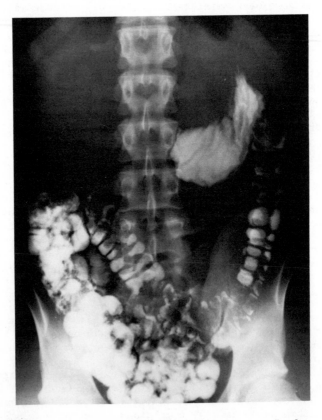

Figure 3. Gastrointestinal series in an "asymptomatic" volunteer forty minutes after oral administration of 225 ml of neat RFC. This subject had a small channel ulcer with pylorospasm and delayed gastric emptying. Note that RFC is still present in the stomach while the left colon is beginning to show filling with RFC.

leaves the food and traverses the GI tract ahead of the food. This behavior of RFC has been attributed to the "creeping" property of substances with low surface tension.

Initially, we thought that rapid gastric emptying might be a disadvantage in certain disease states; however, we found this property to be an advantage in clinical practice. In Figure 3, we see the x-ray of a subject with a small, channel ulcer crater. Gastric emptying of RFC was delayed beyond forty minutes due to pylorospasm of a very modest degree. This subject was a healthy volunteer medical student who stated after the study that he regularly had mild epigastric discomfort especially before sleep and during times of stress. Repeat upper GI series with barium sulfate did not reveal any ulcer crater or delayed gastric emptying.

One disadvantage of RFC in the GI tract of humans is that there is inadequate radiopacification of the esophagus and fundus of the stomach. This inadequacy was not noted in experimental animals.

Indications for Use of RFC in the GI Tract. Obviously, RFC cannot compete in cost with barium sulfate for GI studies. There are specific circumstances in which RFC can and should be used, and these indications represent an estimated one percent of the total market of GI x-ray studies. In any subject where pulmonary aspiration of contrast media is likely, RFC would be useful due to its lack of toxicity to the lungs. These circumstances include infants, elderly patients, and weak cachectic patients as well as all subjects with symptoms of tracheoesophageal fistula. We have also found RFC useful in patients with intestinal fistulas, perforations, or intestinal obstruction, or patients with suspected postoperative ileus or obstruction. The x-ray pictures of the small intestines have been judged superior to those obtained with barium sulfate and can be obtained in less than an hour.

Peritoneal Contamination with Radiopaque Agents

Leakage of contrast material such as barium sulfate into the peritoneal cavity carries potentially severe complications due to the acute and chronic inflammatory reaction incited by barium sulfate. Water-soluble organic iodide compounds may be used when perforation is suspected, but these compounds also produce acute inflammation of the peritoneum. Water-soluble organic iodide compounds produce diarrhea and do not give satisfactory radiopacification of the GI tract. RFC is non-ionic and does not result in diarrhea. RFC has been injected into the peritoneal cavity of experimental animals in dosages of 16 ml./kg., many times the lethal dosage of barium sulfate or oral Hypaque.[R] There was no acute or chronic inflammation even in animals observed for more than four years. No evidence of carcinogenicity was observed on histological studies. The $C_8F_{17}B_R$ leaves the peritoneal cavity very slowly by vaporization and by phagocytosis. Some of the $C_8F_{17}B_R$ was found in subcutaneous lymph nodes of the anterior abdominal wall. Phagocytosis of RFC occurs by monocytes in the peritoneal cavity. Accumulations of vacuole-laden phagocytes can be seen in those areas with radiodensity on x-rays (Figure 4). Large cyst-like accumulations of RFC are seen within the peritoneal cavity. This reaction is comparable to the foreign body reaction seen with Teflon^R or Silastic^R, the most inert materials used in medical implants. Such foreign body reactions are found with any inert material with long residence time.

Fluorocarbons of High Vapor Pressure

The radiopaque fluorocarbon $C_6F_{13}B_R$ was also studied in the peritoneal cavity. This compound has a boiling point of 98° and a vapor pressure of 90 Torr. When injected into the peritoneal cavity, $C_6F_{13}B_R$ vaporizes more rapidly and the abdomen becomes distended with fluorocarbon gas. $C_6F_{13}B_R$ leaves the body more rapidly than $C_8F_{17}B_R$ and is completely eliminated from the peritoneal cavity in weeks even with large doses of 16 ml./kg. This property of rapid vaporization and intracavitary gas formation does not produce any adverse effects in the GI tract or lungs; however, when injected into the subarachnoid space, gas formation results in neurological injury in a significant number of animals. For this reason, we selected $C_8F_{17}B_R$ for our initial investigations of RFC since economic considerations did not permit simultaneous and parallel studies with more than one RFC compound. The $C_6F_{17}B_R$ compound was also non-toxic and may prove to be useful for GI and pulmonary studies. $C_6F_{17}B_R$ vaporizes within the GI tract thus providing useful diagnostic properties of gas contrast. In some manufacturing processes, $C_6F_{17}B_R$ is the principle contaminant or impurity. Up until now, we have held standards of better than 99.9 percent purity for $C_8F_{17}B_R$. This high purity standard may not be necessary, and if not, costs of raw material would be reduced significantly.

Submersion Experiments

Submersion of hamsters in RFC has been performed for periods up to ten minutes. When the hamsters were removed from the liquid, x-rays showed the lungs to be filled with RFC and RFC was also present in the GI tract. Later x-rays showed gradual clearing of the lungs. Since RFC clears primarily by vaporization, the areas of the lungs that are better ventilated show more rapid clearing. Thus, the alveolograms with RFC provide a physiologic picture of different ventilatory patterns in different parts of the lungs.

Microscopic examination of the lungs one day after submersion in RFC revealed the presence of hyperinflation of the alveoli due

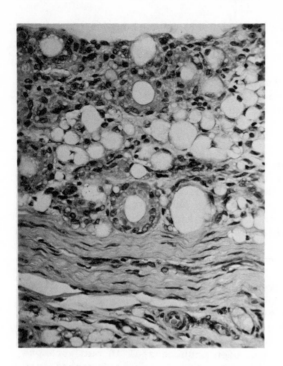

Figure 4. Photomicrograph of the peritoneal surface with an accumulation of vacuole laden phagocytic cells one month after intraperitoneal injection of 16 ml of neat RFC into the peritoneal cavity of a rat. Note the large vacuoles in the monocytic cells at the top of the photograph. Hematoxylin and eosin stain. 175×.

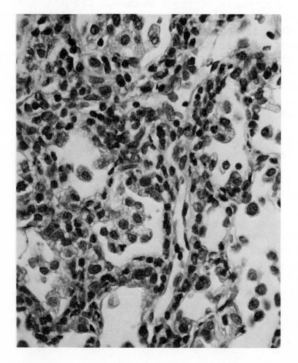

Figure 5. Lung macrophage cells laden with fine vacuoles of radiopaque fluorocarbon in the cytoplasm can be seen occupying the alveoli in a rabbit two days after a bronchogram with 10:1 emulsion of RFC. Hematoxylin and eosin stain. 290×.

to the mechanical effects of the dense fluorocarbon (density 1.93 gm./ml.). Macrophages with vacuoles of RFC were seen in the alveolar spaces (Figure 5). No evidence of pneumonitis due to RFC was observed when compared with control animals. At one month, there were few macrophages with vacuoles of RFC. The macrophages with vacuoles of fluorocarbon showed no evidence of intracellular damage, and the alveolar architecture had returned to normal.

Alveolographic Studies with RFC

X-ray visualization of the alveolar compartment of the lungs cannot be obtained with currently available material. Theoretically, it should be highly desirable to obtain alveolar studies in living subjects without resorting to lung biopsies. Such information should be useful in differentiating various types of lung diseases. In other areas of medicine, precision in diagnosis has been essential in prescribing more effective therapy and in understanding etiologic factors as well as the natural history of disease. At first, we were disappointed to observe that RFC filled the alveolar compartment and did not provide radiopacification of the trachea and bronchi. It then became necessary to develop unique emulsions for tracheobronchography. In time, the alveolographic studies may prove more useful.

Alveolography can be accomplished with doses that are a fraction of the LD_{50} dose. We have been unsuccessful in our attempts to nebulize neat RFC into the lungs. A catheter is inserted into the trachea with topical anesthesia of the larynx and trachea. The neat RFC does not induce coughing or other phenomena of irritation. The topical anesthesia does induce bronchospasm and hypoxemia, but topical anesthesia has been needed for catheter placement in subjects studied thus far. The subjects have complained of pharyngitis and laryngitis from the topical anesthesia and the catheter. We have also observed mild elevations in oral temperatures, white blood cell count, and occasional mild elevations in the serum enzymes SGOT or LDH. These effects have disappeared in twenty-four to forty-eight hours, and it is difficult to determine what part of these adverse effects are due to the catheter placement, the topical anesthesia and the fluorocarbon.

In patients with bullous emphysema, the alveolograms have accurately outlined the areas of normal lung. The emphysematous bullae have not filled with RFC (Figure 6). Areas of centrilobular emphysema have also showed a void on x-rays (Figure 7). Areas of normal alveolar structure were clearly demonstrated in patients and healthy volunteers. Those areas of lung compressed by emphysematous bullae filled slowly on alveolography. Similarly, the areas of poorly ventilated or compressed lung cleared RFC more slowly than did areas of well ventilated lung. Radiographic evidence for the RFC had cleared by forty-eight hours.

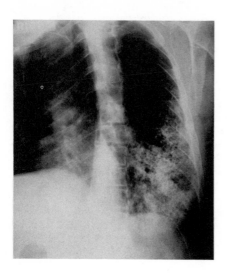

Figure 6. Alveologram with neat RFC in a patient with bullous emphysema occupying the upper portions of the left hemithorax. Note that the fluorocarbon does not fill the emphysematous bullae.

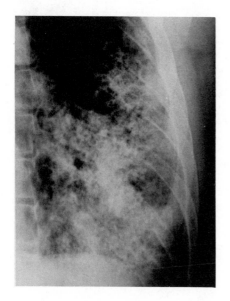

Figure 7. Close up view of the lower portions of the left hemithorax in the same patient as seen in Figure 6. The compressed areas of the lung can be seen to contain radiodense material which is RFC. The punched out areas in the alveologram represent areas of centrilobular emphysema verified by histological examination of the tissue.

Emulsions of RFC

Dilute emulsions of RFC (2:1 volume to volume) are easily
prepared by mixing RFC in a 6% solution of Pluronic F-68 in physiologic salt solutions. Concentrated emulsions (up to 15:1) can
be prepared by gradual addition of neat RFC. The concentrated
emulsions are useful in bronchography. We use emulsions with
concentrations varying from 6:1 to 10:1. Emulsions with concentrations less than 6:1 cause coughing in experimental animals
similar to that produced by physiologic salt solution. Emulsions
of concentrations 6:1 or greater behave more like neat RFC and do
not produce coughing.
The emulsions have a wide range of particle size including
some particles as large as 1 mm. There is some change in particle size distribution as well as viscosity with time, but the
shelf life of the emulsions is very satisfactory for our purposes. Only slight creaming occurs, and the supernatant RFC can be
re-emulsified by shaking the vial. The viscosity of RFC emulsions is non-Newtonian and thixotropic.[5]

Bronchography with RFC Emulsions

Bronchography was performed with topical anesthesia of the
pharynx and trachea and insertion of a tracheal catheter. The
tracheal catheter was positioned into the bronchus of choice, and
10 to 20 ml. of 6:1 or 10:1 emulsion was injected into the main
bronchus with the patient apneic in expiration. The patient was
then asked to take a deep breath. Additional injections into lo-
bar bronchi were performed as indicated. Appropriate x-rays were
taken to obtain complete information on the structure of the tracheobronchial tree. Bilateral bronchograms are usually desired
and have been performed simultaneously without incident. The
x-rays of the chest were obtained with kilovoltage of 70 to 75,
somewhat less than that obtained with conventional x-rays of the
chest.
Satisfactory bronchograms have been obtained in all healthy
volunteers and patients studied. Even patients with severe and
far advanced pulmonary disease have tolerated the RFC emulsions.
Three patients had experienced severe respiratory distress previously when they received bronchograms with currently available
organic iodide bronchographic media. These three patients tolerated the RFC bronchograms well.
Decreases in arterial pO_2 were observed in subjects following topical anesthesia and hypoxemia and bronchospasm have been
reported by others following topical anesthesia of the tracheobronchial tree.[6,7] After the RFC was injected, the arterial pO_2
either decreased or increased or stayed the same. Arterial hypoxemia was corrected by the use of supplemental nasal oxygen
breathing.

Biological Disposition

The biological disposition of RFC has been examined in experimental animals receiving RFC by the several routes studied.
The animals were sacrificed at different intervals after receiving RFC and the tissues were extracted in hexane. The extracts
were analyzed for fluorocarbon using gas liquid chromatography.
The biologic disposition of RFC given by the gastrointestinal
route is shown in Table II. Only those tissues with the highest
fluorocarbon concentrations were listed although other tissues
were also analyzed. Trace amounts of RFC could be seen in x-rays
of the animals twenty-four hours after administration, but chemical analysis revealed significant quantities in the gastrointestinal tract. Trivial amounts of RFC were absorbed from the GI
tract as evidenced by the small quantities of RFC present in
other tissues. A progressive decline in tissue RFC concentration
occurred, and at three weeks there was little or no evidence of
residual RFC.
Tables III and IV present the biologic disposition data on
RFC administered into the lungs for alveolography and for bronchography, respectively. Twenty-one days after administration of
a large dose of 4 ml./kg. of neat RFC, there were only trace
amounts found in the tissues. When RFC emulsion was given in a
dosage of 2 ml./kg., elimination was virtually complete before
twelve weeks after administration.

Lymphography

Radiopacification of lymphatic structures was obtained by
infusion of RFC into lymphatic channels or into lymph nodes.
Neat RFC was preferred for lymphography. Infusion of emulsions
resulted in radiopacification of the lymphatics to the first
lymph node. The emulsion was rapidly phagocytosed by the cells
in the lymph node, the lymph node became engorged and tense, and
the flow out of the lymph node was blocked so that distal lymphatics and lymph nodes could not be visualized. When neat RFC was
infused, the uptake by the lymph nodes was less voracious. The
lymph nodes became radiopaque but not tense, and the distal lymph
nodes and channels were visualized (Figure 8). When excessive
quantities were injected, the neat RFC spilled over into the pulmonary vasculature.
The LD_{50} dose of RFC in cats was between 1.0 and 1.25 ml./kg.
The efficacious dose was 0.2 ml./kg., which was that quantity required to fill the lymphatics before spill over into the venous
system. The residence time of RFC in lymph nodes was greater
than three years in dogs. This long residence time may be an advantage, particularly when one is studying the involvement of
lymph nodes by malignant tumors. Iodized oil, the agent currently used for lymphography also has a long residence time, but this
agent incited inflammation and fibrosis in lymph nodes.

Table II

Biologic Disposition after Gastroenterography

	1 Day	3 Days
Large Intestines	1×10^{-2}	7×10^{-5}
Stomach & Small Intestines	1.4×10^{-3}	5×10^{-6}
Liver	1.6×10^{-4}	6×10^{-5}
Lungs	4×10^{-5}	6×10^{-5}
Lymph Nodes	5×10^{-5}	2×10^{-5}
Fat	5×10^{-5}	1.6×10^{-5}

Tissue levels in ml. $C_8F_{17}B_R$ per gram of tissue at different time intervals after gastroenterography with 16 ml./kg. of $C_8F_{17}B_R$. Experimental animal was the adult rat.

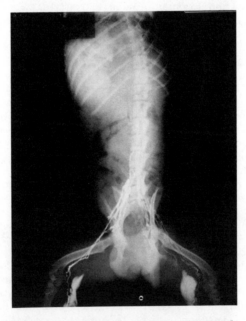

Figure 8. Lymphangiography in a dog with Ethiodal[R] injected into the right leg and neat radiopaque fluorocarbon injected into the left leg. The iodinated oil, Ethiodal[R], is more radiodense than RFC but the difference in radiodensity does not detract from diagnostic accuracy.

Table III

Biologic Disposition after Alveolography

	3 Days	1 Week	3 Weeks
Lungs	8×10^{-4}	8×10^{-5}	5×10^{-5}
Lymph Nodes	7×10^{-5}	4×10^{-4}	9×10^{-6}
Fat	7×10^{-5}	4×10^{-5}	3×10^{-5}

Tissue levels in ml. $C_8F_{17}B_R$ per gram of tissue at different time intervals after alveolography with 4 ml./kg. Experimental animal was the adult rat.

Table IV

Biologic Disposition after Bronchography

	1 Day	3 Days	1 Week	4 Weeks	12 Weeks
Lungs	6×10^{-4}	2×10^{-4}	1×10^{-4}	1×10^{-5}	$< 1 \times 10^{-6}$
Lymph Nodes	2.4×10^{-5}	3.8×10^{-4}	3×10^{-5}	4×10^{-6}	
Fat	7×10^{-5}	9×10^{-5}	1×10^{-5}	4×10^{-5}	$< 2 \times 10^{-6}$

Tissue levels in ml. $C_8F_{17}B_R$ per gram of tissue at different time intervals after bronchography with 2 ml./kg. of a 10:1 emulsion. Experimental animal was the dog.

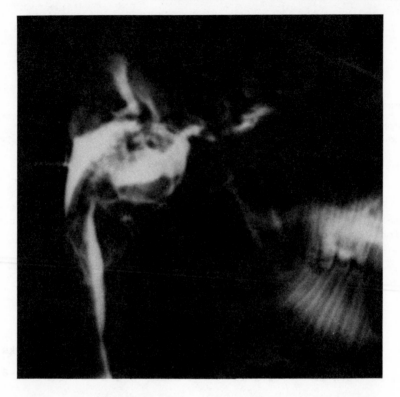

Figure 9. Ventriculomyelogram in a rabbit with neat RFC. The ventriculomyelogram was well tolerated and resulted in excellent visualization of central nervous system structures.

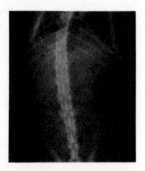 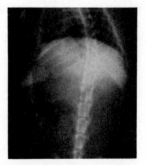

Figure 10. Hepatography and splenography in a rat given 15 ml/kg of a 67% emulsion of radiopaque fluorocarbon intravenously. The particle size of the emulsion was 1–1.5 microns. The x-ray on the left was made prior to injection, and the x-ray on the right was made four days after injection of the emulsion. The structures of the thin spleen are more accurately outlined than those of the more dense liver.

Ventriculomyelography

Radiopaque perfluorocarbon has been used to enhance radiological visualization of central nervous system structures in experimental animals.[8,9] These studies have been performed with $C_8F_{17}B_R$, $C_7F_{15}B_R$ and $C_6F_{13}B_R$. Neat RFC has been used since emulsions were too viscous for this purpose. The RFC compounds proved to be efficacious for outlining appropriate structures of the central nervous system (Figure 9). The residence time of $C_8F_{17}B_R$ in the subarachnoid space was greater than three years. There was no acute inflammatory reaction elicited by $C_8F_{17}B_R$ as evidenced by the lack of significant cellular and chemical changes in the cerebrospinal fluid. No demonstrable neurological injury was produced in experimental animals, either in the acute or chronic phases. A mild arachnoiditis was seen in chronic experiments and was manifested by the accumulation of phagocytic monocytes in the areas of fluorocarbon accumulation.[8,9] Whether or not these subtle changes will be of clinical significance is a moot question. RFC was superior to Pantopaque[R], the currently available contrast agent for myelography.

Perfluorohexylbromide was studied because of the rapid vaporization of this compound at body temperature. The vapor pressure of $C_6F_{13}B_R$ was approximately 90 Torr compared to 14 Torr for $C_8F_{17}B_R$ and 55 Torr for $C_7F_{15}B_R$. Previous studies demonstrated that $C_6F_{13}B_R$ vaporized rapidly and disappeared radiologically within weeks when injected into the peritoneal cavity or subcutaneous space. When injected into the subarachnoid space, $C_6F_{17}B_R$ formed vapor pockets, and the gas phase was not reabsorbed rapidly enough, so that neurological injury and even mortality occurred in animals.[9] The rate of disappearance of $C_7F_{15}B_R$ from the body and the subarachnoid space is under investigation in our laboratory.

Hepatography and Splenography

Intravenous infusions of emulsions of RFC result in radiopacification of the spleen or the spleen and liver. When we tested the bronchographic emulsions for intravenous toxicity, we observed radiopacification of the spleen. These emulsions have a large particle size. The radiopacification was apparent in thirty minutes and increased by four hours. The radiopacification diminished gradually and disappeared by about four weeks. When small particle size emulsions of 1 to 1.5 micron were injected intravenously, radiopacification of the spleen and liver occurred (Figure 10). Emulsions of smaller particle size have been tested with equivocable results. These studies of hepatography and splenography are in an early stage of development so that a considerable amount of research is required to define the optimum particle size and dosage. The RFC is phagocytosed by the reticuloendothelial cells of the liver and spleen (Figure 11). As seen in Figure 11, the radiodensity achieved is sufficient for diagnostic purposes.

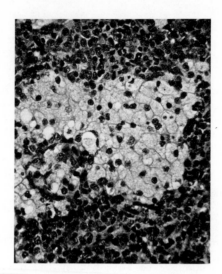

Figure 11. Photomicrograph of the spleen of a rabbit one week after receiving intravenous injection of large particle size emulsion of RFC. The spleen contained foci of mononuclear cells with foamy cytoplasm due to the presence of intracellular fluorocarbon.

Other Areas of Application

Both neat RFC and emulsions of RFC have been tested for arthrography and have been found to be inadequate. RFC has also been evaluated in retrograde urography, cholecystography, and pancreatography in experimental animals. The results obtained were comparable to those obtained with currently available contrast agents. Detailed efficacy and toxicity studies have not been performed to date in these areas of limited frequency of use. RFC may be a feasible alternative for use in patients with hypersensitivity to organic iodide contrast agents. It may also prove to be safer in visualization of inflamed or fragile ductal systems where extravasation of contrast media such as organic iodide compounds would exacerbate the inflammatory process in an organ such as the pancreas.

Summary

Monobrominated perfluoroakyl compounds have been tested as x-ray contrast agents. The compound perfluoroctylbromide is the agent of choice at this time because of its biological inertness, low surface tension, ease of emulsification and favorable rate of elimination from the body. Gastroenterography, bronchography, and alveolography have been performed with radiopaque fluorocarbon in experimental animals and humans. Other areas of application are being investigated in experimental animals.

Acknowledgements: The authors wish to express their gratitude to Miss Pat Lee, Mrs. Frances Multer, and Miss Margo Nielson for excellence in laboratory experiments, and to Doctors Hugh G. Bryce, Raymond J. Seffl, J. Dana McGowen, and Donald La Zerte of the 3M Company.

Supported by grants from the Chemical Division, Minnesota Mining and Manufacturing Company, St. Paul, Minnesota and the U.S.P.H.S. Grant GM20998.

Literature Cited

1. Clark, L.C., Jr., and Gollan, F. Science (1966) 152, p. 1755-56: "Survival of mammals breathing organic liquids equilibrated with oxygen at atmospheric pressure."
2. Clark, L.C., Jr., Kaplan, S., Becattini, F. J. Thoracic Cardiovascular Surg. (1970) 60, p. 757-773: "The physiology of synthetic blood."
3. Geyer, R.P. Fed. Proc. (1975) 34, p. 1499-1505: "'Bloodless' rats through the use of artificial blood substitutes."
4. Long, D.M., Liu, M., Szanto, P.S., Alrenga, D.P., Patel, M.M., Rios, M.V., and Nyhus, L.M. Radiology (1972) 105, p. 323-332: "Efficacy and toxicity studies with radiopaque perfluorocarbon."
5. Arambulo, A.S., Liu, M., Rosen, A.L., Dobben, G., and Long, D.M. Drug Devel. Commun. (1975) 1, p. 73-87: "Perfluoroctyl bromide emulsions as radiopaque media."
6. Salisbury, B.G., Metzgir, L.F., Altrose, M.D., Stanley, N.N., and Cherniak, N.S. Am. Rev. Respir. Dis. (1974) 109, p. 691: "Effect of fiberoptic bronchoscopy on respiratory performance in patients with chronic obstructive pulmonary disease."
7. Miller, W.C. and Awe, R. Am. Rev. Respir. Dis. (1975) 111, p. 739-741: "Effect of nebulized lidocaine on reactive airways."
8. Dobben, G.D., Long, D.M., Szanto, P.S., Mategrano, V.C., and Liu, M. Neuroradiology (1973) 6, p. 17-19: "Experimental studies with radiopaque fluorocarbon in the subarachnoid space."
9. Brahme, F., Sovak, M., Powel, H., and Long, D.M. Acta Radiol. Scan. In Press: "Perfluorocarbon bromides as contrast agents in radiography of the central nervous system."

Discussion

Q. How can the bromoperfluorocarbons compete with barium sulfate in studies of the GI tract?

Dr. Long

A. Perfluoroctyl bromide and the others cannot compete with $BaSO_4$ on a cost basis. However, for selected cases, barium can be dangerous. For about 1% of the GI studies done, water soluble contrast agents are used at the present time. These contrast agents are not well tolerated, causing diarrhea and irritation, and don't give very good x-ray contrast. The fluorocarbon is vastly superior. The fluorocarbon is superior to barium in the small intestines also. There are other ways where the barium isn't as adequate as the fluorocarbon. For reasons that are not clear, we get good coating of the esophagus and other parts of the stomach in dogs with fluorocarbons, but not in humans. In humans, we have never been able to get good coating of the esophagus or the fundus of the stomach, so that's a limitation. Radiologists and surgeons are very enthusiastic about these fluorocarbons in the GI tract, although there will be an increased cost.

Q. Would you comment on the C_7 compound?

A. We have tried the C_7 compound and it is very interesting. It is not eliminated rapidly enough from the body to be useful in myelography. For alveolography, you can use the C_6, C_7 or C_8 compounds. C_6 may be better than the others. For clinical studies, we have used the C_8 compound.

Q. What about irritation?

A. There is little or no irritation from this material.

Q. You mentioned in the case of myelography how the fluorocarbons are retained for a long period of time. Have you tried to relate that to Pantopaque retention?

A. Pantopaque is retained permanently. You don't remove all that agent. The fluorocarbon is more acceptable than Pantopaque since it does not cause arachnoiditis like Pantopaque does. There are water soluble contrast agents that are being used now in myelography, and we are not sure how these are going to work out. In some of the earlier studies, they did not appear to be toxic when first injected. Even though they completely disappeared, an arachnoiditis or chronic inflammatory process was incited by the initial injection of the water soluble contrast agents. The non-ionic water soluble contrast agents are now being evaluated clinically to see whether they will be more acceptable, and the results have been encouraging. The fluorocarbon is vastly superior to Pantopaque, the material currently used in almost all myelography.

Q. I am surprised that you do not feel more favorable about the use of fluoroca rbons in the specific application of myelography.

A. I was not comparing fluorocarbons with Pantopaque, but with the non-ionic water soluble contrast agent under study. I think this contrast agent is preferred, but it is not approved by the F.D.A. It is being used in other countries, and it is under clinical investigation in the U.S. It looks like the non-ionic contrast agent is going to be a very good agent but fluorocarbons will be able to compete with it under some circumstances, such as in patients with hypersensitivity to iodinated compounds.

10

Preparation and Physiological Evaluation of Some New Fluorinated Volatile Anesthetics

DONALD D. DENSON, EDWARD T. UYENO, ROBERT L. SIMON, JR., and
HOWARD M. PETERS

Stanford Research Institute, 333 Ravenswood Ave., Menlo Park, Calif. 94025

Although several adequate fluorinated anesthetics are in
clinical use today, all have disadvantages and possible hazards.
We are synthesizing and evaluating fluorinated ethers for use as
volatile anesthetics. Since these compounds are "inert" gases,
they exert their biologic effects without undergoing any chemical
transformation during administration, residence in the body, and
elimination from the body. It is hoped they will provide the
advantages of currently available fluorinated anesthetics but
preclude their disadvantages (1).

In discussing volatile anesthetics, it is important to understand the difference between analgesia, narcosis, and anesthesia. Analgesia (Stage I) is the loss of pain or numbing of
sensory nerves without loss of consciousness. Narcosis (Stage
II) is a reversible state of analgesia accompanied by stupor or
unconsciousness. Surgical anesthesia (Stage III) is the reversible loss of all modalities of sensation and loss of consciousness. The planes of surgical anesthesia in humans (2) are:

Plane 1: Swallowing reflex lost; respiration regular;
          muscle relaxation minimal
Plane 2: Muscle relaxation increased
Plane 3: Muscle relaxation further increased and
          suitable for intraabdominal surgery
Plane 4: Skeletal muscle relaxation complete;
          possible cyanosis; blood and pulse pressure
          falls; pulse rate increases

The level below Plane 4 of surgical anesthesia is respiratory
arrest (Stage IV). While planes 1-3 of surgical anesthesia are
the most important, some consideration must be given to the

dangerous aspects of Stage III, Plane 4, and Stage IV. These
levels must be carefully avoided if the patient is to survive
without adverse affects.

Thus the major objectives of successful anesthesia (2) are
to:

(1) Alleviate pain by blocking sensory or afferent nerves.
(2) Block mentation to alleviate the mental anguish and
      anxiety resulting from the fear of pain.
(3) Relax the muscles by blocking efferent or motor nerves.
(4) Preclude adverse effects of surgery or anesthesia.

Examination of the fluorinated anesthetics in use today
suggests that fluorinated ethers (as a class) offer more advantages than fluorinated hydrocarbons. While these ethers are
more unpredictable in terms of biological activity, they have
three important advantages: (1) they provide a long transition
between analgesia and anesthesia, (2) they provide more effective muscle relaxation, and (3) they demonstrate less sensitiza-
tion of myocardial tissue to epinephrine.

Hazards of Fluorinated Anesthetics Now in Use

Halogenated compounds have been used as anesthetic agents
since 1847 with the discovery of chloroform. During the 1940s,
Robbins reported the first comprehensive study of fluorocarbons
as potential anesthetics (3).

While any number of fluorinated compounds have potential
anesthetic properties, wide scale clinical use has been limited
to about eight compounds. The four compounds used most often
are:

Halothane	$CF_3CHBrCl$
Methoxyflurane	$CH_3OCF_2CHCl_2$
Fluroxene	$CF_3CH_2OCH{=}CH_2$
Enflurane	CHF_2OCF_2CHClF

All these compounds are associated with acute and chronic toxic-
ities that must be carefully considered (1, 4).

Halothane is oxidatively metabolized to trifluoroacetic
acid (5).

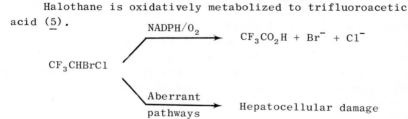

This transformation is probably harmless with production of few,
if any toxic intermediates. It is quite possible however that
aberrant metabolic pathways produce reactive intermediates capable of inflicting hepatocellular damage (1). Halothane is
implicated in producing an unpredictable postanesthetic hepatitis
(1), and in some cases sensitizes myocardial tissue to epinephrine (6).

Methoxyflurane produces free fluoride ion on metabolism (5).

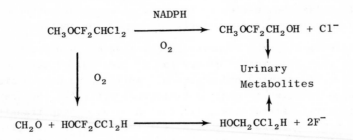

In fact since 60-80% of all absorbed methoxyflurane is metabolized, relatively high serum fluoride levels can be expected.
These levels are often high enough to cause renal tubular cell
damage, giving rise to the high output renal failure syndrome (4).
This nephrotoxicity is further complicated by the observation
that phenobarbital may sensitize a threefold or greater extent of
methoxyflurane metabolism to fluoride ion (4).

Fluroxene is metabolized in animals to trifluoroethanol and
in humans to trifluoracetic acid (5).

$$CF_3CH_2OCH = CH_2 \rightarrow CF_3CO_2H + \text{Urinary Metabolites}$$

High serum fluorides and heptocellular damage are associated with
its use in humans. In addition fluroxene is extremely flammable,
and it is primarily for this reason that it has been removed
from clinical use.

Enflurane is tranformed to unknown metabolities and fluoride
ion in humans (5).

$$CF_2HOCF_2CHClF \rightarrow \text{Unknown Metabolites} + F^-$$

Although these fluoride ion levels are routinely low, in some
patients they reach 80 μ m/l, which is dangerously close to the
normal fluoride tolerance level (2).

Chronic exposures to traces of these fluorinated anesthetics
can be a hazard. According to a recent ASA study (7), operating

room personnel have shown an increase in spontaneous abortion,
malformation of children, cancer in female anesthesiologists,
liver disease, and kidney disease.

Synthesis of Fluorinated Ethers

While only a few fluorinated ethers are in actual clinical
application, others have shown promise. An example of a class
of cyclic diether anesthetics is 4,5-dihalo-2,2-(bis)trifluoro-
methyl-1,3-dioxolanes, 1.

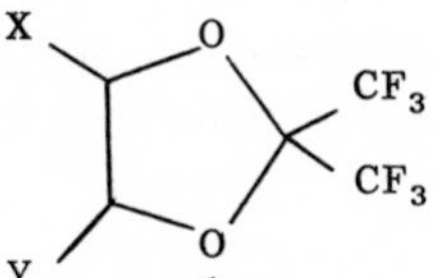

Gilbert (8) patented the use of the parent 2,2-(bis)trifluoro-
methyl-1,3-dioxolane 1 (X = Y = H) as an inhalation anesthetic
in 1967. Terrell and Moore (9) then patented the use of 4,5-
dihalosubstituted-2,2-(bis)trifluoromethyl-1,3-dioxolanes 1 in
1973.

Gilbert found his material to be more potent than halothane
(8). While no definitive Anesthetic Index data are provided by
Terrell et al., it appears that the halosubstituted materials
are more potent than the parent compound (9); 1,3-dioxolane
itself has limited anesthetic properties (10).

Our experience in the conversion of carbonates to difluoroformals led us to investigate the possibilities of preparing
equally as potent, but more stable, cyclic diether inhalation
anesthetics. Our approach is based on the following reaction:

$$R-O-\overset{\overset{\displaystyle O}{\|}}{C}-O-R' \;+\; SF_4 \quad \xrightarrow[\Delta]{Cat.} \quad R-O-CF_2-O-R'$$

Because of the ready availability of the starting material, the
ease of preparing a host of halogenated analogs, the apparent

success of the 4,5-dihalo-2,2-(bis)trifluoromethyl-1,3-dioxolanes
as anesthetics (9), we initiated our investigation using ethylene
carbonate, 2.

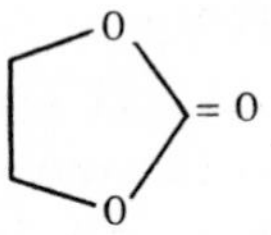

2

We extended the conversion of carbonates to difluoroformals to
include the preparation of 4,5-dihalo-2,2-difluoro-1,3-dioxolanes:

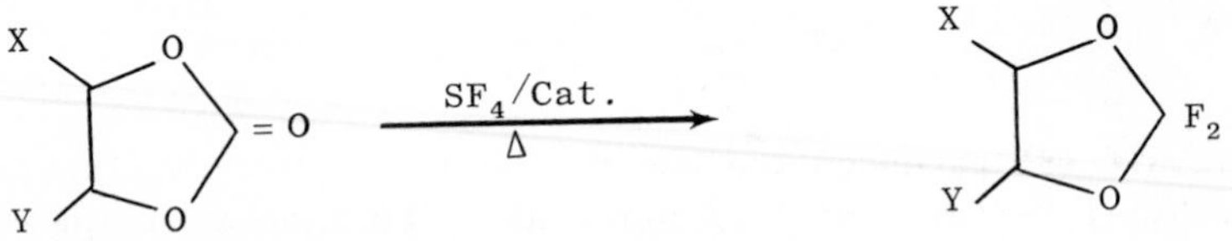

where X and Y are halogens

   Ethylene carbonate, 2, can be chlorinated to give 4-chloro,
3, and 4,5-dichloroethylene carbonate, 4.

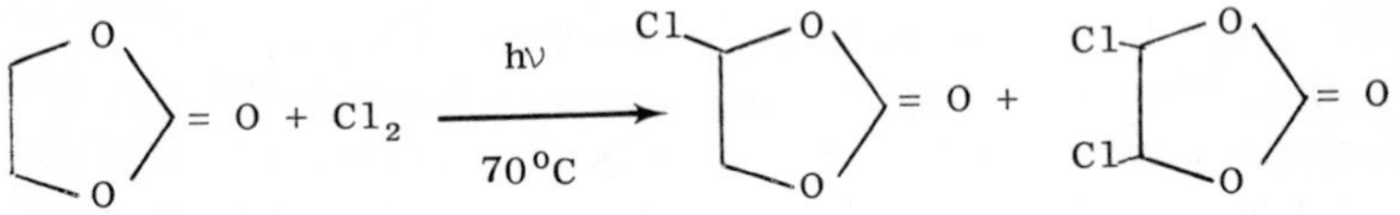

2   3   4

Monochloroethylene carbonate, 3, can be dehydrohalogenated to
vinylidene carbonate, 5, as shown below (11).

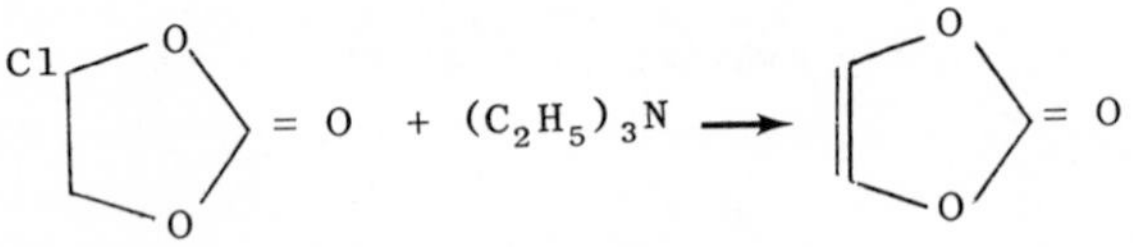

3   5

Vinylidene carbonate, 5, can be converted to a number of other
halogenated derivatives as shown in Scheme 1.

Compounds 6, 7, 8, and 9 are readily obtained by the use of
polyhydrogenfluoride/pyridine reagent, as described by Olah et al.
(12). Straightforward hydrobromination or bromination (11)
affords compounds 10 and 11. A similar series of derivatives
can be prepared from the dehydrohalogenation of 4,5-dichloro-
ethylene carbonate, 4.

   The fluorination reactions of these carbonates have led to
much exciting and useful chemistry. The fluorination of 4,5-
dichloroethylene carbonate, 4, for example, gives three products:

The product mixture ratio is extremely dependent on the cata-
lyst/SF_4 ratio. An increase of $\geq$ 0.5 in the catalyst/SF_4 ratio
results in the formation of 4-chloro-2,2,5-trifluoro-1,3-
dioxolane, 13, as the major product. The chlorinated-unsaturated
1,3-dioxolane, 14, suggests that the formation of 4-chloro-2,2,5-
trifluoro-1,3-dioxolane, 13, results from a dehydrochlorination
reaction followed by a hydrogen fluoride addition reaction. The
unsaturated-1,3-dioxolane, 14 is an important finding in this
reaction as we will see in the fluorination of 4-chloroethylene
carbonate, 3.

Because we observed halogen exchange in the fluorination of
4,5-dichloroethylene carbonate, 4, we began our investigation of
the monochlorinated derivative 3 using an HF/SF_4 ratio of 0.45.
The totally unexpected result was the formation of 4-chloro-
2,2,5-trifluoro-1,3-dioxolane, 13, as the only isolated product.

For this compound to form, a hydrogen substitution reaction must
be occurring. There is no reported evidence for either a direct
halogen exchange or a dehydrochlorinationhydrogen fluoride addition reaction. This result is independent of cat/SF_4 concentration and occurs whether HF or TiF_4 is the catalyst. Heating the
precursor carbonate with either HF or TiF_4 at 150°C for 24 hr
results in no reaction. Heating 4-chloroethylene carbonate, 3,
with SF_4 in the absence of catalyst for 24 hr results in recovery
of the starting carbonate, 3. Hydrogen substitution by SF_4 has
been reported by Applequist and Searle (13).

Experiments conducted at lower temperatures resulted in
either no reaction or in the formation of 2,2,5-trifluoro-1,3-
dioxolane, 15, as the major product.

A similar result was obtained in attempts to fluorinate
ethylene carbonate, 2. The objective of this experiment was to
prepare 2,2-difluoro-1,3-dioxolane for comparison with 1,3-
dioxolane to determine the contribution of the difluoroformal
moiety to biological activity. In these experiments, the HF/SF_4
ratio was again maintained below 0.5. The formation of 2,2,4,5-
tetrafluoro-1,3-dioxolane, 16, was independent of both temperature and catalyst concentration.

The hydrogen substitution reaction observed for the fluorination
of 4-chloroethylene carbonate, 3, is the predominant reaction in
this case. We have been unsuccessful in all our attempts to prepare 2,2-difluoro-1,3-dioxolane. This reaction is under further
investigation to determine the feasibility of preparing this
compound.

Structure-Activity Relationships

The major structure-activity relationships for volatile
anesthetics are as follows:

- Halogenation of hydrocarbons and ethers increase
  in potency in the order $I > Br > Cl > F$.
- Unsaturation increases potency.
- Fluorine addition decreases potency, boiling
  point and flammability and increases stability
  of adjacent halogen atoms.
- Increased potency in a homologous series follows
  an increase in molecular weight, boiling point,
  and oil/gas coefficient.
- One or more hydrogen atoms are necessary for CNS
  depression.

Since molecules containing only carbon, fluorine, hydrogen, and
oxygen are not usually very potent anesthetics, we attempted to
introduce one chlorine into 2,2,4,5-tetrafluoro-1,3-dioxolane,
16. These experiments are summarized in the following equations.

Photochemical chlorination of 16 with one equivalent of chlorine
at room temperature results in the formation of 4-chloro-2,2,4,5-
tetrafluoro-1,3-dioxolane, 17, along with five products. The
maximum yield of this product, 17, obtained to date is 53%. Further chlorination of the reaction mixture results in an increase
in the by-products, with no significant increase in the desired
4-chloro-2,2,4,5-tetrafluoro-1,3-dioxolane, 17.

Photochemical chlorination of 2,2,4,5-tetrafluoro-1,3-di-
oxolane, 16, with excess chlorine results in the formation of
4,4,5,5,-tetrachloro-2,2-difluoro-1,3-dioxolane, 18, as the major

product (>95%). This reaction is under further study and will
be reported in a future publication. We do not believe that
4,4,5,5-tetrachloro-2,2-difluoro-1,3-dioxolane, 18, will be a
potent anesthetic since there are no hydrogen atoms in the molecule. Generally one or more hydrogens are required for CNS
depression.

The structure-activity relationships presented above suggest
that a more potent member of this class would contain a bromine
atom. We have attempted to maximize anesthetic potency without
the incorporation of bromine, because it is a thermodynamically
weak bond in metabolic environments. However in an effort to
determine how well our class of diethers adheres to the reported
structure reactivity relationships given above, we attempted to
fluorinate 4,5-dibromoethylene carbonate, 11. The major product
from this reaction is 4-bromo-2,2,5-trifluoro-1,3-dioxolane,
19.

11   19

As in the case of the fluorination of 4,5-dichloroethylene
carbonate, 3, we have observed halogen exchange, but here this
conversion is independent of cat/SF$_4$ concentration. We have not
been able to isolate the unsaturated bromine containing 2,2-
difluoro-1,3-dioxolane, 20.

20

Experiments with TiF$_4$ at 100° and 150°C resulted in no
exchange in the absence of SF$_4$. Similarly, experiments with SF$_4$
in the absence of catalyst resulted in no exchange. Additional
experiments involving the fluorination of both 4,5-dibromo-
ethylene carbonate, 11, and 4-bromo-5-fluoroethylene carbonate,
6, are in progress.

Physiological Evaluation

The potential anesthetic agents synthesized in this research were tested in mice by procedures similar to those
described by Burgison (14). A wide-mouth, screw cap, glass jar
was flushed with oxygen for one minute, and a measured amount of
a synthesized compound in a syringe was ejected from a syringe
onto the bottom of the jar. Initially the amount of a test substance that would saturate the "air" in the jar was calculated
according to the methods reported by Ough and Stone (15).
Several concentrations of each test compound that would not
saturate the "air" were evaluated. The jar was closed quickly
and evaporation of the substance was facilitated by gentle
rotation of the container. Five mice were quickly dropped into
the jar, and the bottle was capped immediately.

Every 15 seconds, the container was gently rotated and the
time required for each animal to become anesthetized (loss of
righting reflex) was noted. Since the occurrence of induction
after 0.5 min but no later than 5 min is considered to represent
the optimal induction time for known potent anesthetic agents,
the interval between these two time limits was used as a crite-
rion of induction.

The mice were kept in the jar for 10 minutes, then removed
and placed on a pan. The time of recovery (to righting) was
noted for each animal. Postanesthetic analgesia was tested by
pinching the base of each animal's tail every five minutes. A
squeak on pressing the tail was considered as the end point. If
an aminal still showed analgesic response a half an hour after
being removed from the jar, the pinch test was administered
every 15 minutes. The subjects were kept for twenty-four hours
to record any latent toxicity of the compounds.

The evaluation of the substances in mice was conducted in
two stages. In the first stage, a dose-range experiment was
performed to determine the lethal concentration range, and three
appropriate concentrations were selected. Fifteen animals were
tested at each of the three concentrations. In the second stage,
another dose-range study was conducted to chose three lower concentrations. Fifteen animals were tested at each selected
concentration to determine the medium effective anesthetic
concentration. The anesthetic potency and margin of safety of
each compound were expressed as an anesthetic index.

The median anesthetic concentration (AC_{50}) is defined as an
estimated concentration by which 50% of the animals are expected
to be anesthetized. The median lethal concentration (LC_{50}) is

an estimated concentration by which 50% of the animals are expected to die. The anesthetic index (AI) is the ratio of LC_{50} to AC_{50}. The larger the anesthetic index the greater margin of safety. The lower the AC_{50}, the lower the concentration of drug available for metabolism. However, a drug acting at lower concentration is not necessarily safer. As already discussed, methoxyflurane is nephrotoxic, yet its AC_{50} is about half that for halothane.

Before presenting our physiological data, it is worthwhile reviewing some data available for other anesthetics. These data are summarized in Table 1 (2, 8). While no hepatocellular or nephrotoxicity data are available for 2,2-(bis)trifluoromethyl-1,3-dioxolane, it is interesting to note that its AI is slightly larger than halothane (8). This molecule contains only carbon, fluorine, hydrogen, and oxygen. While no definitive AI data are reported by Terrell (9) for the halogenated analogs, it is noteworthy that there was very little increase in activity.

Table 1

ANESTHETIC POTENCY FOR FLUORINATED ANESTHETICS

Anesthetic	B.P. (^{o}C)	AC_{50}	LC_{50}	AI	MAC
Halothane	50	0.78	2.74	3.50	0.78
Methoxyflurane	105	0.3	-	-	0.23
Fluroxene	43	1.2-8	-	-	6.0
Enflurane	57	2-4	-	-	2.2
2,2 (Bis)trifluoromethyl-1,3-dioxolane	100	0.5	2.38	4.70	-

We began our physiological evaluations using halothane as a standard. In these experiments we obtained an AI of 4.75, which is slightly higher than the AI of 3.50 previously reported (see Table 1). At anesthetic concentrations, halothane is characterized during induction by ptosis accompanied by lacrimation. Recovery was characterized by "flat tails." Once the animal regained the righting reflex, they maintained it. In addition, rapid recovery times were observed, indicating rapid halothane elimination. (Rapid elimination is often associated with lower lipophilicity.) At lethal concentrations, dark eyes and blanched ears were observed in addition to ptosis and lacrimation. Table 2 summarizes our findings for halothane.

Table 2

EFFECTS OF HALOTHANE CF_3-CHBrCl, ON MICE
(AI = 4.75)

Conc., Vol %	No of Animals	Induced before 30 sec	Induced before 5 min	Mean Recovery Time, min	Deaths
0.94	15	-	5	0.78	0
1.00	15	-	7	0.83	0
1.06	15	-	12	1.03	0
4.00	15	8	15	7.72	3
5.00	15	13	15	8.13	9
5.50	15	14	15	8.83	12

At anesthetic concentrations of 4,5-dichloro-2,2-difluoro-1,3-
dioxolane, 12, minimum hind leg movement and ptosis were observed,
but no lacrimation or convulsive behavior. Recovery was characterized by the animals running in circles. Mice were wobbly and
had a bouncy gait. Recovery times for 12 were much longer than
for halothane, as shown in Table 3, suggesting that, even at low
concentrations, elimination is not nearly as rapid as it is for
halothane. Slower elimination suggests a higher lipophilicity
for 4,5-dichloro-2,2-difluoro-1,3-dioxolane, 12. Lethal concentrations were characterized by irregular respiration and apparent
bradycardia. Ptosis and lacrimation were observed in a few
animals.

Table 3

EFFECTS OF 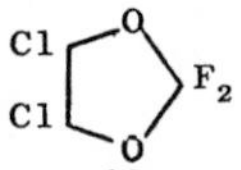 ON MICE

12

(AI = 7.88)

Conc., Vol %	No of Animals	Induced before 30 sec	Induced before 5 min	Mean Recovery time, min	Deaths
0.25	15	-	0	-	0
0.30	15	-	2	0.63	0
0.35	15	-	13	10.60	0
2.00	15	-	15	60.08	0
2.50	15	-	15	60.47	6
3.00	15	1	15	120	14

Anesthetic concentrations of 4-chloro-2,2,5-trifluoro-1,3-dioxolane, 13, were characterized by hypoactivity, slight convulsiveness, ptosis, and occasional lacrimation. During recovery the animals often ran in circles and some lost their righting reflexes. Recovery times, given in Table 4, were about half those reported for 4,5-dichloro-2,2-difluoro-1,3-dioxolane, 12 (see Table 3). Although the AI determined for 4-chloro-2,2,5-trifluoro-1,3-dioxolane, 13, was approximately the same as the AI determined for halothane, recovery times were somewhat longer for 13. At lethal concentrations, the animals' eyes became extremely dark (exophthalmos) and deep jerky respiratory response was followed by shallow breathing and bradycardia.

Table 4

EFFECTS OF (structure) F_2 ON MICE

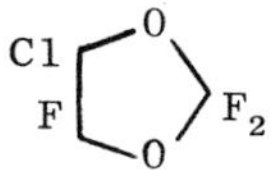

13

(AI = 4.66)

Conc., Vol %	No. of Animals	Induced before 30 sec	Induced before 5 min	Mean Recovery time, min	Deaths
0.88	15	–	5	1.95	0
0.94	15	–	13	3.07	0
1.00	15	–	14	4.13	0
3.00	15	–	15	15.48	0
4.00	20	–	20	17.78	10
5.00	15	5	15	19.72	12

The preliminary evaluation of 2,2,4,5-tetrafluoro-1,3-dioxolane, 16, is summarized in Table 5. Rapid hind leg movement ptosis and tachypnea were associated with anesthetic concentrations. During recovery, the animals ran in circles, and some shivered for a few minutes. Recovery times were far more rapid than those observed for either 4,5-dichloro-2,2-difluoro-1,3-dioxolane, 12, or 4-chloro-2,2,5-trifluoro-1,3-dioxolane, 13. During induction, lethal concentrations of 2,2,4,5-tetrafluoro-1,3-dioxolane were characterized by the same observations made for

the anesthetic concentrations. During recovery, however, hind
legs were extended and dragged, and animals crawled with their
front legs.

Table 5

EFFECTS OF $\begin{smallmatrix} F \\ F \end{smallmatrix}\left[\begin{smallmatrix} O \\ O \end{smallmatrix}\right\rangle F_2$ ON MICE

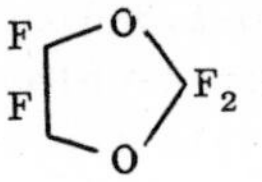

16

Conc., Vol %	No. of Animals	Induced before 5 min	Mean Recovery Time, min	Deaths
4.0	10	2	0.92	0
5.0	5	2	1.42	0
6.0	5	5	2.37	0
10.0	5	5	5.58	3
12.0	5	5	-	5

Statistical Analysis

The data obtained from the administration of the lower concentrations of halothane were analyzed by calculating the percentage of animals that were anesthetized at each of the three
lower concentrations. The percentages were plotted against concentration on logarithmic probability paper as shown in Figure 1.
Similarly, the percentage of mice anesthetized at each of the
three lower concentrations of 4,5-dichloro-2,2-difluoro-1,3-
dioxolane, 12, and the percentage induced at each of the three
lower concentrations of 4-chloro-2,2,5-trifluoro-1,3-dioxolane,
13, were plotted. From the graph, the AC_{50} of each compound was
computed according to the method of Litchfield and Wilcoxon (16)
as shown in Table 6. Since AC_{50} of 4,5-dichloro-2,2-difluoro-
1,3-dioxolane, 12, is significantly lower than that of halothane,
the former is considered to be more potent than the latter in
inducing anesthesia.

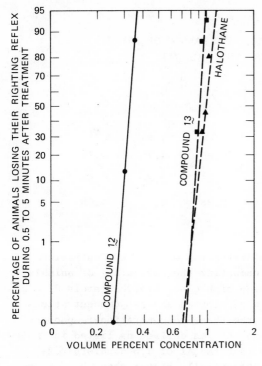

Figure 1. *Effects of inhalation anesthetics on mice*

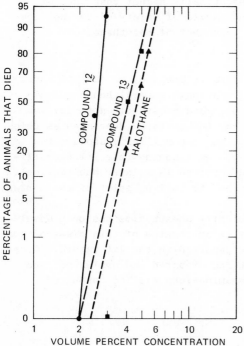

Figure 2. *Effects of high doses of inhalation anesthetics on mice*

Table 6

ANESTHETIC POTENCY AND MARGIN OF SAFETY
STRUCTURE VS. ACTIVITY

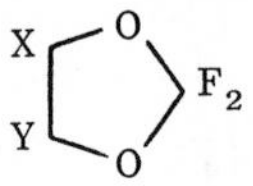

Compound	X	Y	AC_{50}	LC_{50}	AI
Halothane			0.99	4.7	4.75
12	Cl	Cl	0.33	2.6	7.88
13	Cl	F	0.88	4.1	4.66
16	F	F	-	-	-

The results of the administration of high concentrations of compounds were analyzed by computing the percentage of animals that died at each of the three high concentration levels of each compound. The percentages were plotted against concentrations as shown in Figure 2. From the three concentration-response curves, the LC_{50} of each compound was derived as listed in Table 6. Further analysis showed that the AI of 4,5-dichloro-2,2-difluoro-1,3-dioxolane is substantially higher than that of halothane, indicating that the margin of safety of the new compound is considerably higher than that of halothane.

Summary

From the summary of our data in Table 6, we can see some trends in structure-activity relationships. Although our dioxolanes do not contain bromine, good anesthetic activity is present. As could be predicted, the activity decreases as chlorine is replaced by fluorine. The lipophilicity appears to decrease as chlorine is replaced by fluorine (partition coefficients will be determined and reported in a later publication). Anesthetic activity increases as the boiling point of the series increases.

A new class of potent volatile anesthetics has been developed. The structure activity relationships of this class of compounds appear to follow those reported in the literature. Other compounds in this class are being prepared and evaluated, and more detailed physiological examinations will be conducted.

Acknowledgements

The authors gratefully acknowledge the support of this research by Institutes of Health, General Medical Sciences Division, under Grant Number 5-R01-GM-20082-02.

Discussion

Q: How long did you observe the animals for evidence of latent toxicity?

A: In all cases for at least 24 hours.

Q: Are the compounds stable to base?

A: Yes! You can convert the difluoroformal group back to the carbonate using concentrated sulfuric acid. These compounds are also stable to mild acid.

Q: Is there any splitting out of carbonyl fluoride?

A: No, at least we have not observed this.

Q: The possibility exists for other isomers. Can you comment on that?

A: We cannot predict whether we are studying one active isomer or dealing with unknown mixtures. We can say that if other isomers are present, we are not observing an acute lethal effect.

Literature Cited

1. Cascorbi, H. F., "Anesthesia Toxicity," in 1974 Annual Refresher Course Lectures, American Society of Anesthesiologists Annual Meeting, Washington, D. C., October 12-16, 1974, Lecture number 227 and references cited therein.
2. Larsen, E. R., "Fluorine Compounds in Anesthesiology," in Fluorine Chemistry Review, P. Tarrant, Vol. 3, pp. 1-44 (1969) and references cited therein.
3. Robbins, J. H., J. Pharmacol. Exptl. Therap. (1946), 86 197.
4. Brown, B. R., Jr., "Enzymes and Anesthesia," in 1974 Annual Refresher Course Lectures, American Society of Anesthesiologists Annual Meeting, Washington, D.C., October 12-16, 1974, Lecture Number 225 and references cited therein.
5. Van Dyke, R. A. and Chenoweth, M. B., Anesthesiology (1965), 26, 348.
6. Tucker, W. K., Rackstein, A. D., and Munson, E. S. Brit. J. Anaesth., (1974) 46, 392.
7. Cohen, E., et al., Anesthesiology (1974), 41, 321.
8. Gilbert, E. E., (to Allied Chemical Co.), U.S. Pat. 3,314,850 (1967).

9. Terrell, R. C., and Moore, G. L., (to Airco, Inc.), U.S. Pat. 3,749,791 (1973).

10. Virtue, R. W., Proc. Soc. Exptl. Biol. Med., (1950), 73, 259.

11. (a) Newman, M. S., and Addor, R. W., Amer. Chem. Soc. (1953) 75 1263.
    (b) Newman, M. S., and Addor, R. W., J. Amer. Chem. Soc., (1955), 77, 3789.

12. Olah, G., Nojima, M., and Kerekes, I., Synthesis, (1973) 779.

13. Applequist, D. E., and Searle, R., J. Org. Chem. (1964), 29, 987.

14. Burgison, R. M., "Animal Techniques for Evaluating Anesthetic Drug," in Animal and Clinical Pharmacologic Techniques in Drug Evaluation, J. H. Nodine and P. E. Siegler, eds., Yearbook Medical Publishers Inc, Chicago, Ill. (1964), pp. 369-372.

15. Ough, C. S., and Stone, H., J. Food Sci., (1961), 26, 452.

16. Litchfield, J. T., Jr., and Wilcoxon, F. A., J. Pharm. Exptl. Ther. (1949), 96, 99.

INDEX

INDEX

211